A Better Way–

Manufacturing Techniques for the Leading Woodworking Companies of Tomorrow

by
Terry Acord

Chartwell Communications
Des Plaines, Illinois

A Better Way —
Manufacturing Techniques for the Leading Woodworking
Companies of Tomorrow

Copyright ©1999, Chartwell Communications

Published by:
  FDM Magazine
  Chartwell Communications
  380 E. Northwest Highway, Suite 300
  Des Plaines, IL 60016
  847/390-6700

Printed in the United States of America
ISBN 1-57450-050-3

Introduction

The publication of this book represents the achievement of a professional goal I've had for more than six years. Since the late 1980s I have wanted to write a book that could be used by both people on the shop floor and by managers in changing the way they run their businesses.

In that time I have met many managers and shop floor workers who, like me, struggled with the changes that were taking place in our industry. I felt at various times frustrated, stupid, overwhelmed, and mystified at what I perceived as my failure to contribute to the success my employers wanted to achieve. To understand the depth of my desire to make a difference, it might help to know something of my background.

I grew up in the coal mining town of Lorado, West Virginia, surrounded by the working class. Almost everyone I knew was a miner or earned their living from the mining industry. I even worked in the mines to help pay for college. This instilled in me a deep respect for blue-collar (when you work in the coal mines, black-collar) workers; their desire to feel pride about the work they do; and the need for a sense of accomplishment.

I attended West Virginia University, in the heart of the "steel country" that surrounded Pittsburgh. I fully expected to work in the metalworking industry when I graduated.

This was all changed, of course, by the economic upheaval that occurred in the U.S. during the 1970s and 1980s. Seemingly overnight, the metalworking industry found itself in a bitter war for survival. The decline of the American automobile industry had hit steel manufacturers hard. Even harder to handle, however, were new Asian competitors, who somehow could buy ore here in the U.S., ship it to Japan, process it, transport it back overseas to the U.S., and still sell it cheaper than the Americans.

So it happened that I went south to interview, and ended up in the furniture industry. Although I had no background in furniture, I soon fell in love with wood-working. There's still something magical to me about taking a piece of wood and transforming it into beautiful furniture.

During my career I watched the companies I worked for struggle, move ahead, and drop back. All the while I felt I was riding the waves on a ship without a sail. I was helpless to aid in steering us in the direction we wanted to go. It might have been easier if I could have pointed to apathy on the part of myself, the workers, management, or owners, but I couldn't. I knew all these people and felt they all wanted the company to succeed.

When I learned about J-I-T techniques, and read the works of Hall, Schonberger, and others, I began to understand why the Japanese had been able to overtake us in so many industries. I began to see that the problem with American industry in general, and the furniture industry in particular, was not one of attitude or knowledge, but one of philosophy.

At first I couldn't really believe it was that basic, that simple. I couldn't accept that ideas alone could determine whether a company or an entire industry fails or succeeds. But consider it for a moment. Whatever company you work at, whatever product or service you make, didn't it find its birth as someone's idea? Wasn't the industrial "golden age" the result of the ideas of people such as Eli Whitney, Henry Ford, Fredrick Taylor, and Frank and Lillian Gilbreth?

Although I slowly came to believe that the problems American Industry faced stemmed from a basic change in how industry operated, I also could see the obvious advantages that Japanese concepts provided. Yet, I still didn't feel they were the whole solution. They were closer to the center of the target, but I still didn't feel they were a "bull's eye."

I found my "bull's eye" in 1985, when I read a book called *The Goal*. It had been written by an Israeli physicist name Eliyahu Goldratt, who had the boldness to claim, in so many words, that he knew what was wrong with modern industry. The book itself, co-authored by Jeff Cox, was unusual. It wasn't written as a textbook, it was in the form of a novel — some even called it a love story.

I took the book home one Friday and finished it around 3 a.m. the following Sunday. It was as if I had literally been struck by a lightning bolt. I distinctly remember saying, "That's our plant!" time and time again. I now know from talking to other readers that this reaction is the norm — not the exception.

I walked into the office Monday morning, ready to transform our plant. Anyone who has ever come back from a seminar charged up and ready to implement the new ideas they had learned can probably guess at what happened next. I went and talked to the manager who had given me the book.

Yes, he had found it equally interesting. Yes, he thought it applied to our plant. Yes, he agreed that it could potentially have a huge impact on how our company was run. Set up a task force to look at implementing it? Heavens, no!

We had our annual physical inventory coming up, and the usual end of the month push for production was on. He said we should plan on lunch next week and talk about how we might put together a proposal to try some of the ideas on a small scale, first. Then he asked if the reports I had been working on were ready yet.

Needless to say, the lunch never took place, the task force was never formed, and the implementation was never started, much less finished. Over the years I continued playing the role of fire fighter, working under whatever philosophy the company that paid my salary followed. Don't get me wrong, some of these were successful companies, and had achieved good success with their manufacturing systems. Still, I knew we weren't achieving all we could, we weren't working toward "the goal" as well as we might.

This is not unusual. To date *The Goal* has sold over 1.5 million copies. Practically every American company, as well as many overseas, have been exposed to its concepts. Today the ideas are taught under the various headings of "Theory of Constraints," "Synchronous Manufacturing," "Flow Manufacturing," and others. Yet relatively few of the companies exposed to the concepts have actually set off on the course described. Why? If it's so intuitively correct, so logical, why do people resist it?

I think the answer is simple because it involves change.

Change is never comfortable, and changes as fundamental as those needed for flow manufacturing mean that you will encounter a lot of resistance to their implementation.

So how do you overcome this resistance? The first step is knowledge and understanding. Everyone must understand the need for change, the advantages of change, and the means of change. Next, they want to know the rewards of change, what's in it for them? Third, the champions of change need the fortitude to keep the company on course during the painful transition period. It's my hope that some of those needs will be met with this book.

I have tried to show some of the basic steps needed to make the transition to a flow manufacturing environment. For that reason this book is actually two books in one. The odd-numbered chapters are in novel form, similar to *The Goal*. This format makes for a much easier read for those who don't have an interest in the technical aspects of synchronous manufacturing. Even-numbered chapters give more of the "bread and butter" details of implementation, and are intended for those who will be directly responsible for making it happen.

Although this book is set in the woodworking industry, I feel confident that if you work in a different industry, the methods and mindset presented will work equally well.

One final thought. In the end the adaptation of a new manufacturing philosophy, no matter how logical or well presented, is a leap of faith. You don't know that these concepts will work for your company until you try them, and that means going into it based on your belief and nothing else. I hope this book will help firm your convictions and steel your resolve. The concepts are right; the ideas do work; there is a better way. Take a deep breath, and jump in!

Acknowledgments

If I have seen further it is by standing on the shoulders of Giants.
— Sir Isaac Newton

Most of the content of this book is not totally new concepts; it's a representation of ideas that have come from the true innovators in the area of manufacturing improvement.

First, I want to thank the authors of *The Goal* — Dr. Eliyahu Goldratt and Jeff Cox. Eli is the father of the concepts that have grown into Synchronous Manufacturing, and he and Jeff gave us a novel that is still changing the professional lives of managers everywhere.

In addition, I want to thank the members of the company I work with — the MPI Group. Many of the "war stories" contained in this book are derived from the experiences of the MPI Associates, who have more than 100 years of collective experience in companies throughout North America.

A special thanks should go to Al Podzunas, principal of the MPI group, who contributed a bulk of the "war stories" and the foreword. Al has been at the forefront of implementing these concepts in American industry and is one of the most experienced people in the world in the field implementation of Synchronous Manufacturing.

Finally, I want to thank the good people at Cahners Publishing who printed the initial series of articles in *FDM/Furniture Design & Manufacturing* magazine that the book is based around. Special thanks to Bruce Plantz, editor of *FDM*, who worked with me to publish the original set of articles; Michael Chazin, associate publisher at Cahners, who backed the idea of transforming the articles into a book; and finally to Jean Hyland, managing editor at *FDM*, who patiently pointed out my errors and needed rewrites for the article series and the book. Their patience and support made this a truly rewarding and enjoyable experience.

Dedication

This book is dedicated to:

The "Gamma Group"
Pete, John, Jeff, Michael
"Friends of a lifetime"

Mom, Dad, Larry, Sherry, and Debbie
"For believing"

And most of all,
My wife Trudy and sons Quentin and Wesley
"The reason for it all"

Preface

Many books are written about how to change the culture, how to improve the process, and how to make money in business today. Most are theoretical, or tell you how someone else accomplished it. But none give you the detail on how you can do it yourself. This book gives you the "Meat & Potatoes" way to get started.

A few words about the author, Terry Acord. Today risk takers or street fighters are few and hard to find in manufacturing. It is even harder to find both in one person.

When I first met Terry, he reminded me of Luke Skywalker in the Star Wars Trilogy, and I was Yoda to Terry's Luke Skywalker. Luke wanted to learn the ways of the Force and be a Jedi warrior to beat the dark side; Terry wanted to learn why American manufacturing was losing its edge (the Force) and how to change the tide of what was happening to American industry.

Terry was a great student and carries on the burning desire to change and save American jobs and make America "the Force" in manufacturing again. I hope as you read this book you feel the Force and want to be a Jedi warrior in your company, plant, or corporation.

Albert E. Podzunas, Jr.
Principal
MPI Group
Wallingord, CT

Contents

Chapter 1

The Hero Is Coming

It was lunchtime in the break room at Average Furniture Company, Oaktown, USA. Al, Benny, and Carl, all supervisors for Average, were playing cards and talking shop. Al and Benny were 20-year veterans of the company. Carl, an industrial arts graduate, was a recent addition who had just finished Average's management trainee program.

Benny pulled a thick, folded stack of papers from his back pocket and threw it on the table in disgust.

"Cripes!" said Benny. "They changed next week's production schedule again! How can they expect us to plan anything when they do crap like that?"

"I'll tell ya," Al replied. "I've come to the conclusion that the guys up in 'The Tower' don't have any idea what our customers are buying."

"It definitely ain't like the old days," agreed Benny.

"What do you mean?" asked Carl. "Are you saying it didn't used to be this crazy?"

Benny shook his head. "Nah, you should'a been here 10 years ago. Back then we'd only have to cut a series of furniture once or twice a year. Why, we'd be running our biggest selling suites for two or three months straight."

"Well, jeez, why can't we do that anymore?"

"It's the dang designers," Benny exclaimed. "They can't

design furniture that sells in any quantity any more. It used to be we never had more than five or six suites in the line at one time. Nowadays we have to keep 10 or 15 suites just to keep the same volume as back then."

"I'll tell you who else is doing it," chimed in Al. "It's the schedulers. Used to be we didn't even start a cutting until it was oversold. Nowadays they got us making things even though there's a six-month supply in the warehouse. Meanwhile, we got stuff that's sold that isn't even on the schedule yet."

Al turned to Carl. "You run the packing and shipping department," he said. "How often have you had to work overtime to get out shipments at the end of the month, only to have to go on short time the next week because you don't have any work?"

Al and Benny went on to name several more guilty parties: the government, foreign manufacturers, greedy plant owners and stockholders, engineers, accountants. Carl was feeling increasingly uneasy. "If the answer is so obvious," he thought, "why isn't someone fixing the problems?"

A Rotten Hand

Carl sat in the break room long after Al and Benny had left. His cup of vending machine coffee grew cold in his hand. He didn't even like coffee, really. He just drank it because everyone else drank it. He stared at the playing cards that circled the paper cup. He had been dealt a pretty rotten hand, he noticed.

Carl hadn't really intended to go into the furniture business. He had grown up in northern West Virginia, in the heart of the steel belt between Pittsburgh, and Cleveland. He had always assumed he would return there after college.

But that was before the Japanese and Europeans had taken over the steel industry. By the time he graduated, the steel companies were laying off left and right. It didn't seem like a good industry to be in, even if he had been able to find a job.

Besides, he had fallen in love with the Carolinas. When the InterMast corporate recruiter visited the campus, he signed up for an interview. Carl soon found himself a management trainee at Average Furniture.

No, he thought, it definitely wasn't where he had pictured himself, but that didn't mean he didn't want to do a good job.

He had known in school that he wasn't really learning to be a manager — that he had only been "learning to learn," so to speak. The trouble was, he had been a management trainee for six months before taking the packout/shipping supervisor's job, and he still didn't know anything.

Oh, he'd learned to fill out time card reports, efficiency reports, production summaries, vacation forms, and other assorted paperwork. He'd learned how to pump his people up when the crunch was on (usually the end of the month) and how to tell them they were on short time (usually the beginning of the month). He had hired and fired workers, and handled one case of substance abuse, one case of domestic abuse, and even one fight in the department.

But Carl still felt like a referee, calling the fouls but not affecting the game. Whatever reasons Average had for hiring him, it couldn't be just to maintain the status quo. He always assumed his job was to make it better, to improve things somehow.

He had done pretty well in school. He had always worked hard in the plant and done well at whatever he had been shown to do. "I'm not stupid," he thought. "Why do I feel like such a failure?"

Filling Dad's Shoes

In another part of the plant, Frank Average, Jr. was sitting at his desk, asking himself the very same question.

Frank's father had started the Average Furniture Company nearly 40 years before. His father and uncle had been synonymous with the company for most of its existence, along with a handful of loyal employees. Uncle Hal had handled the finances, and Dad had done everything else — design, engineering, production, and sales. And he had done them all superbly.

Frank Sr. had taken a business with the unlikely name of "Average Furniture Company" and turned it into an above-average success. He had been a true renaissance man. He could spend all day putting together a killer marketing campaign, and all night charming new customers. It had meant a lot of nights away from home, but he had built a company to be proud of.

Frank Jr. always admired his dad. Other kids might rebel against following in their fathers' footsteps, but not Frank Jr. He had always known he would join his father in the furniture industry. "Yep," his father would say, "once the sawdust gets in your veins, you're hooked."

Filling Dad's shoes, however, turned out to be a lot harder than Frank Jr. had expected. It wasn't that he couldn't do all of the things his father could do. He could — he just wasn't quite as good at them as Dad.

And even Frank Sr. had had a hard time holding things together the last 10 years. It seemed business was getting tougher and more competitive every day. Customers wanted and expected more every year. Even long-time customers started to drift away to bigger suppliers that offered more variety, better service, and lower prices. Try as Average might, the company couldn't seem to keep up with the big manufacturers.

After a long discussion, Dad and Uncle Hal finally decided to sell. The InterMast Corp., a big furniture conglomerate, had expressed interest repeatedly. Eventually, Frank Jr. watched his father sign the papers that turned his fate over to InterMast. It was one of the hardest days of Frank Jr.'s life.

Frank Sr. had retired soon after. Part of the sales agreement was that Frank Jr. would assume the role of plant manager, with his father acting as an assistant during the transition. InterMast had made the usual statements about keeping the management team in place, wanting to build on past success, and so on.

Even so, the acquisition meant learning a new way of doing things, and Frank Jr. suspected his dad just couldn't stand having someone else tell him how to run his company.

A Cutthroat World

In a lot of ways, Frank Jr. felt he had failed his father. Dad disagreed. "I couldn't run the business successfully in today's world, son," he said. "Business is a lot more cutthroat, a lot meaner, nowadays. It seems it's just not enough to be good at what you do. It definitely doesn't seem like as much fun as it used to be."

Then, Frank Sr. had placed his hand on his son's shoulder. "I wish I had some answers for you," he said.

"Answers," Frank Average thought. (He had dropped the "Jr." about a year after his dad had retired). InterMast had told him to get his efficiencies up, and he had raised them almost 20 percent. They told him to get his work center utilization up, and he had raised it almost 10 percent. They told him to reduce his standard costs, and he had initiated a massive cost-reduction program that had taken an average of 4 percent out of each product.

Every time InterMast had made a demand, Frank had answered. But through it all, the plant hadn't made a net profit, and gross profit had actually slipped a bit.

The quarterly review had been this morning. After the usual mind-numbing recitation of facts and figures, the folks from corporate had dropped their bomb. They were sending down a consultant, they said, to help Frank find the answers he needed to run his plant profitably. He was to give the consultant his full cooperation in all areas.

Frank knew what this meant in the corporate world. They didn't think he could cut it as a plant manager. They were sending this guy down as a last-ditch effort to make a good manager out of him. More than likely, the consultant had also been instructed to evaluate Frank's abilities and to report on the possible need to replace him.

Now, sitting in his office, Frank watched the workers filing out of the plant to go home. Some ran with more energy than they had shown all day. Some strolled casually, talking with long-time friends from the plant. Frank couldn't help a sarcastic thought.

"Don't worry, folks. The man with all the answers will be here soon. Our hero is coming!"

Chapter 2
New Rules for Manufacturing

Practically everyone would agree that business today is a lot tougher, faster paced, and more competitive than in years past. Has manufacturing really changed? If so, when did it change? What made it change? And most important, how should we respond to these changes?

It's worthwhile to ask those questions, because they will help us understand the forces that shaped today's manufacturing environment.

The 'Good Old Days'

To find answers, we need to look back at those "good old days" of manufacturing. Early in this century, Henry Ford introduced mass production to America and used it to bring the automobile within reach of the average American. The trade-off for the average car buyers was little or no control over how the car looked, what options it came with, or any other product-related element. Every version of the Model A and Model T was made one way and one way only.

When World War II came, mass production methods and something called statistical quality control played pivotal roles in helping American industry meet the staggering demand for war goods. American workers were able to build more and higher quality arms, supplies, vehicles, and everything else needed for the war effort. This manufacturing capability was a key factor in our ability to win the war.

The years following the end of W.W.II were indeed something of a golden age for American manufacturing. Much of the European and Asian industrial base was in ruins after the war. Because most American consumers had foregone replacing basic household items during the war years, the mighty American manufacturing machine that had been churning out radar and tanks shifted over to refrigerators and televisions. The world was rebuilding, and America was its sole supplier.

The Production Mindset

In the race to meet this global demand, American manufacturers placed less emphasis on quality and more on increasing production. The key measurements were efficiency (the time to perform an operation compared with some established standard), standard cost (a summation of labor and material costs), and machine utilization (the hours a machine is run divided by the total hours it is available).

The key feature to remember is that these measurements were internal. They measured the plant's performance against how the plant operators thought they should be doing. The measurements didn't take into account the relationship between the plant and its customers.

In the post-war era, factories didn't need to worry about customers. Demand was high. Manufacturers could pretty much sell anything they could get out the door. Even in the late 1960s, American manufacturers still owned the American consumer. Although European and Asian countries had rebuilt their manufacturing capacities, American buyers generally associated their products with inferior quality.

The New Global Consumer

The turning point came with the OPEC oil embargo in the early 1970s. Overnight, America found itself at the mercy of the oil-producing nations of the Middle East and hostage to its own excessive appetite for petroleum products.

During this period, Americans began to take another look at the products available from other countries, particularly automobiles. Many foreign manufacturers already made cars with higher fuel efficiencies than their American counterparts. Volkswagen Beetles, Honda Civics, and Toyota Corollas began showing up in large numbers.

Changes in the attitudes of American consumers were even more fundamental. Consumers began to realize they had choices. The change didn't happen overnight, but more and more consumers were less willing to "Buy American" simply out of some vague sense of patriotism. They wanted better products, more options, and better service. Other countries, especially Japan, were eager to meet those needs.

American managers, most of whom had grown up and spent their years in the glory days of manufacturing, suddenly found themselves with a new breed of customers. Customers who no longer were willing simply to buy what they had on hand. Customers who would not accept shoddy quality. Customers who would not accept long waits for their products. And, if you couldn't meet their needs, they were more willing than ever to switch to someone else who could.

Think about workers in the automotive industry. All at once, they were losing customers to their Japanese counterparts. Their managers began demanding better quality, better productivity. They began to see the number of jobs in their industry dwindle, and all the while they were being driven to raise their level of performance.

At night, though, these workers would go home, take off their hard hats and uniforms, put on their blue jeans, and head to the mall. They had suddenly become the consumers of microwaves, cordless phones, and, yes, furniture. When it came time to part with some of their hard-earned money, don't you think their demands as consumers would be influenced by what they went through in the workplace?

Once this "new consumer" snowball started down the hill, it continued to gain in speed and size. A lot of American companies were flattened by it, although they never saw it coming.

The manufacturing game had changed forever, and managers were going to have to scramble hard to learn how to play by a very different set of rules.

Chapter 3

The Profitability Puzzle

Carl walked through the offices of the Average Furniture Company, lost in thought. He was still disturbed by the conversation he had had with Al and Benny about how the plant had changed in the last few years.

On impulse, he stopped in at the office of the plant scheduler, Donnie. Donnie had been in production for several years before going into scheduling. Maybe he could shed some light on the problem, or at least give Carl a different perspective.

Carl found Donnie in his usual spot, buried behind his desk full of printouts, charts, and reports. He knocked lightly. "Got a minute, Donnie?"

Donnie gave a rueful smile. "Like my Granddad used to say, 'Once the water rises past your nose, it don't matter how deep it is.' Come on in, Carl, what can I do for you?"

Carl gave a brief outline of last week's talk with Al and Benny. "So how about it, Donnie? Has scheduling changed a lot over the past few years?"

Donnie paused, rubbing his chin. "The short answer to that is, 'Yes.' Like most things, however, it isn't that simple.

"Al and Benny are right when they say that our cuttings are a lot smaller than they were 15 years ago. They're also right about having a lot more suites to provide the same volume as back then. However, if they want to blame it on the designers, then every designer in our industry must have caught the 'stupid disease' at once, because I don't know of any companies that aren't going through the same thing.

"I also understand their frustration with scheduling. I know it must seem like we just jerk them around, especially at the end of the month. However, they have to realize that production brings a lot of that on themselves."

Donnie pulled two charts from the pile on his desk. "Look, here's a copy of some charts I do for a 'Conformance to Budget' report every quarter. As you know, we offer our customers a one-month fixed lead time. In other words, if a customer places an order on May 1, we commit to ship by June 1.

"This first chart is the customer orders by week from April, May, and June. The second chart is the shipments by week from May, June, and July. Given our one-month lead time, you would expect the second chart to be exactly like the first, right — that April's orders would match May's shipments exactly?"

"Allowing for some minor deviations, yeah, I'll buy that," said Carl.

"Well, look at what happens in real life." Carl looked at the two charts.

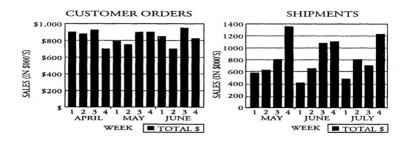

"But they aren't alike at all!"

"That's right," said Donnie. "Even though they average about the same total dollars per week, the shipments don't reflect customer orders. What's more, we have a similar pattern every month, no matter what our orders do. We always ship light at the beginning of the month and heavy at the end."

"But why?"

Donnie smiled. "You're in charge of packing and shipping, buddy. You tell me."

Carl didn't return the smile. He left, feeling more confused and lost than ever.

An Unassuming Beginning

Frank Average paced as he waited at the airport gate. The corporate office had been very tight-lipped about this consultant guy — assuming it was a guy. Frank only knew his name, Wes Quinten, and that he would be arriving on this flight. Calls to other InterMast plant managers were unenlightening. Either Quinten had not visited any of their plants, or they weren't telling.

Frank sighed. He would know soon enough. He watched the plane taxi to a stop. He had started to make a cardboard sign with Quinten's name on it, but he always felt like an idiot holding one of those things. At the last minute, he had borrowed a jacket with the company logo on it. It was hotter than heck, but it beat holding a sign.

As passengers from the flight started filing out, he looked at each face, trying to figure out which one was Quinten. Just what did a know-it-all look like, anyway? The travelers filed past, and Frank was beginning to get that worried feeling you get when you think you have missed someone at the airport, when he felt a touch at his elbow.

"Mr. Average? I'm Wes Quinten."

Frank turned to see a man in a warm-up suit, carrying a garment bag and briefcase. He had seen him get off but hadn't really looked at him; he was expecting someone in a suit. Frank took Quinten's hand as he re-examined his corporate savior. Middle-aged, slightly under 6 feet tall, a light graying of the hair around the temples. In a suit and tie he would look like a thousand other guys that walked through this airport every day — certainly no white knight.

"Maybe I was expecting armor and a shield," Frank thought wryly. To the consultant he said, "Happy to meet you, Mr. Quinten. Welcome to Oaktown. We're glad to have you with us."

Quinten smiled slightly, as though he didn't quite believe Frank. "Call me Wes, please. We're going to be spending too much time together to bother with titles. Is it all right if I call you Frank?"

"Fine with me. Well, let's head down to baggage claim to get the rest of your things, then we'll get you to your hotel. We put you up at the ShadeTree Manor. It's right across the street from the plant. You could walk to work if you wanted."

"That's fine. If you don't mind, I would like to get started as soon as I can get checked in."

"Sure," said Frank. "Do you want to take a plant tour this afternoon?"

Wes smiled. "Actually, I was wondering if there was a pool hall near the plant."

A Whole New Game

Frank hadn't been in Red's Billiard Box in almost eight years. When he had started taking a hand in running the plant, he decided it was time to project a little more con-

servative image. Besides, a lot of the plant people liked to
come down here in the evenings, and he didn't want to
intrude on their off hours. Folks needed a place they
could get together and complain about the boss, and that
was hard to do when the boss was shooting pool on the
next table.

At two in the afternoon, though, there weren't many
people in Red's. A few of the regulars, some of whom
were retirees from Average, clustered around the television
in the corner. Frank waved at Red and pointed at the least
worn table in the hall. Red nodded and took a container
of balls and a square of cue chalk off the counter and carried them over.

"Hey, FJ, haven't seen you in ages. How's your dad?"

Frank winced at the nickname but said, "He's doing
great, Red. Still pretending to fish and golf while he's
drinking beer."

Red laughed. "Must be nice, I don't think I'll ever get to
retire, as little as I make off this place. When you talk to
him, tell him I said he still has five bucks of mine that I
want a chance to win back."

"I'll tell him."

Red placed the balls and chalk on the table. With a nod
at Wes, he returned to his place behind the counter. Frank
watched him go. Maybe old Red didn't make a mint off
this place, but he didn't have many problems to deal with,
either. In a lot of ways Frank envied Red. There were sure
worse ways to live.

The sound of balls clacking interrupted his musings.
Wes had dropped the balls on the table and was racking
them. Frank walked over to the cue rack and began exam-
ining the sticks, trying to find one that was moderately

15

straight and had a tip on it.

Over his shoulder he spoke to Wes. "I have to admit, Wes, you've surprised me so far. I expected you to show up in a three-piece, ready to get down to business. Instead, you show up looking like you just came off your morning walk, and drag me down to a pool hall first thing."

Wes laughed. "That's a reaction I'm used to getting. I gave up wearing suits a long time ago; I've found it makes me and the people on the floor more comfortable. As for the pool hall, I think it's extremely important that we get a chance to know each other and lay our cards out on the table, before we have to deal with supervisor, employee, and production issues. I find that a pool table is preferable to a conference table when you want people to feel comfortable about speaking their minds."

Wes hung the triangle on the lamp that swayed over the pool table. "For example, all pleasantries aside, you can't be real thrilled with me being here. You have to feel that it's some sort of indictment against your management abilities, right?"

Frank looked Wes in the eye. "Yeah, I was thinking something along those lines."

"I don't blame you. I would feel the same. That's one of the things I wanted to talk to you about. Frank, it's not a mark against you that I came here. In fact, it says a lot about what the corporate managers think of your abilities that they chose your plant for my project."

Wes walked over to the cue rack, trying to find a good stick. "I had a chance to review the records on most of the plants at InterMast. One thing that stood out in your case is your ability to respond to corporate directives. When they told you to get a measurement up, you got it up. You always responded with positive results."

Frank shook his head. "Yeah, too bad it didn't result in the plant making more money."

The consultant stopped chalking his cue. "You consider that important, do you?"

Frank had been lining up to break the balls; now he jerked back up. "Important? Of course it's important! The reason the plant is here is to make money. That's why I don't understand why improving all those measurements didn't help." He broke the balls with a sharp crack. He got a lot of action, but no balls went down.

Wes considered his shot for a long time. "Maybe the reason is that those measurements don't have a darn thing to do with whether or not your company makes money." He pocketed the three ball. "I've got solids," he said as they moved around the table.

Frank could hardly believe what he had heard. "What do you mean, they don't have anything to do with it? If you want to make money, you have to run an efficient operation, you have to utilize your people and machines well, you have to have low standard costs. I mean, everyone knows that!"

Wes was smiling and nodding as he lined up his next shot. He pocketed the six ball, then said, "Tell you what, how about a little bet?"

"Sure, now that you already have two balls down, you want to bet! OK, I'm game. What's the bet?"

Wes walked over to a small blackboard hanging on the wall, normally used for playing darts. "It is your contention that efficiency, utilization, and standard costs have a direct relationship to profitability, correct?" He wrote on the blackboard:

$$\uparrow \text{ EFFICIENCY } = \uparrow \text{ \$PROFITS?}$$
$$\uparrow \text{ UTILIZATION } = \uparrow \text{ \$PROFITS?}$$
$$\downarrow \text{ STD. COST } = \uparrow \text{ \$PROFITS?}$$

"Everybody knows that," said Frank. "It's common sense."

"OK, here's the bet. If you win, you get dinner for two on me. If I win, you have to spend the evening writing the equations that relate each of these three factors to profitability."

Frank frowned. "What do you mean?"

Wes pointed to the blackboard. "If, as you say, these factors directly affect a company's profitability, then you should be able to write a mathematical relationship for each of them, right? An equation that directly ties each measurement to profitability. Something like 'Efficiency times X minus Y equals profitability.'

"So what do you say? Do we have a bet?"

Frank shrugged. "I get the distinct feeling I am being hustled here, but, sure — what the heck!"

A Search for Answers

That night, Frank settled in on his couch with a pad and pen. He had been hustled, all right. Wes had only missed once during their entire game. Frank hadn't played in eight years, and it showed. So now he was going to pay his part of the bet by trying to write the formulas as promised.

In front of him were several of his old college textbooks — management accounting, operations research, and so on. He had already gone through them, hoping to find the equations written out. He would have loved to have been able to just photocopy the appropriate sheets

out of a textbook and hand them to Wes. Unfortunately,
he had not been able to find the equations anywhere.

Oh well, it still shouldn't be too difficult. Let's start
with an easy one: efficiency. "The more efficient an operation is, the more profitable it is." Frank wrote at the top of
his pad. Well, in general that was true, but it didn't give
him a precise mathematical relationship.

Something nagged at him. He got his briefcase and
pulled out a mound of papers and reports, one of which
was the "Conformance to Budget" report Donnie had
shown Carl earlier that day. After a few minutes of shuf-
fling, he had two sets of reports arranged on his coffee
table. One showed the plant efficiency by week; the other
showed shipment dollars. Since the company shipped
everything direct to the customers, he felt this was a
pretty good measure of profitability.

Frank drew a grid on his pad and calculated the average
efficiency by week for the previous 12 months; he then
did the same for dollar shipments. Then he averaged week
#1 of each month, week #2 of each month, and so on, and
wrote the results on his grid:

Week of the Month	% Efficiency	Shipment $
1st Week	72	$480K
2nd Week	84	$640K
3rd Week	89	$800K
4th Week	68	$1,200K

Frank went back and checked the individual weeks. He

found the same thing Donnie had mentioned: Shipments followed a consistent pattern of low at the beginning of the month and high at the end. Efficiency, on the other hand, tended to peak in the middle of the month and go down the last week and first week of each cycle. Was there some kind of time lag here? No, the pattern for shipments remained regular even when the efficiency didn't follow its pattern. He wasn't going to be able to prove a direct connection with these numbers!

He had even less luck with utilization. The plant utilization figures were calculated on a monthly basis, and these numbers didn't give have any clear relationship to profitability that he could see.

With a sigh, he turned to the one factor he was sure he could tie into a formula. He quickly wrote:

$$\textit{Profit} = \textit{Selling Price} - \textit{Standard Cost}$$

"There," he thought. "Wes will have to accept that formula; there's no arguing it."

But he heard himself arguing. If this formula were true, why were there "variance accounts?" And why were there "standard costs" and "actual costs?" If a standard cost wasn't the actual cost, what the hell was it?

Actually, he knew. It was a marker. At the beginning of each fiscal year, the accountants froze all the costs at that point in time. They used that cost list as a benchmark for the remainder of the year. So the further into the year you went, the wider the variation between standard and actual costs.

Even worse, Frank knew that some of the assumptions that went into the standard cost calculations were pure bull. For example, the standard size of a load used to allocate setup times was 400 pieces. Ha! Frank couldn't

remember the last time he had actually had a chance to
run 400 pieces. Anytime he complained about it, however,
the answer was, "Get us an accurate sampling and we will
change it." As if he had time to have his production
people run around and take a survey!

Frank sighed. In good conscience he couldn't even
defend the standard cost equation. It looked like he was
going to fail Quinten's 'test,' whatever it was. What was it
Wes had said, almost offhand? "Maybe the reason is that
those measurements don't have a darn thing to do with
whether or not your company makes money."

"Lord," thought Frank, "practically everything we do is
geared toward those measurements. If we aren't doing it
to make money, why are we doing it?"

Chapter 4

The Rules Have Changed

Today's manufacturing managers find themselves in a "game" that fundamentally changed in the early 1970s with the emergence of a new type of consumer. It changed from being a seller's market, driven by internal measures, to a buyer's market, where companies have to compete for customers.

Many, in fact most, companies failed to change their dynamics to meet the demands of this new "game." By continuing to run by the old measurements, companies pay a price in today's environment.

The Old Rules

The old measurements used to evaluate a plant were:

1. Efficiency, measured as **actual time** to do a task/engineering standard time for that task.

2. Machine utilization, measured as **time a machine is run/total time available** on that machine. Note: Utilization also can refer to work stations, departments, even whole plants.

3. Standard cost: A summation of the labor and material costs for each part of a product. The labor cost was usually multiplied by some factor to estimate overhead costs for each part.

There also were standard profitability measurements (net profit, return-on-investment, and cash flow), but these were the domain of the accountants and administration. Production people below the level of plant manager rarely became involved in these considerations because it was difficult to relate them to specific activities on the plant floor.

Old vs. New

Let's examine how running under the old measurements helps or hurts us under the new rules. We'll start with an example that deals with efficiency.

When a chair plant that I worked at some years ago started offering customers fixed lead times, we found ourselves with a problem. The assembly department complained that parts were not arriving fast enough. They would set up to run a chair and, invariably, one or two parts would be missing, which prevented them from building the chairs.

We did time studies, verified the EOQ's (economic order quantities), and ran various other analyses. Aside from determining that the parts missing were usually 4/4 upholstery parts, we could not figure out what was causing the problem.

It wasn't until we started spending a lot of time just walking through the mill, observing the operation, that the answer became clear. Most of the interior parts of the upholstery frames were 4/4 upholstery-grade wood. The exterior of the chairs were usually 6/4 and 8/4 higher grade lumber. These were the process steps for each type of material.

Exposed lumber	**Upholstery lumber**
1. Double-end trim and bore	1. Band saw
2. Shape	2. Double-end trim (old machine)
3. Sand	3. Vertical boring

As we would walk through the machine mill on Monday, we would see the double-end miter saws and shapers covered up with work, while the machines that ran the upholstery parts sat idle. Near the end of the week, the situation would reverse itself. The double-end trim and bore machines would be out of work and the upholstery part machines were covered up. To find out why, we went to the rough mill supervisor.

The rough mill's efficiency was measured on two factors: board feet cut per man hour and lumber yield. It was assumed that having both numbers high meant the plant was doing well financially, at least as far as the rough mill was concerned.

The problem was obvious. To drive those two measurements as high as possible, the foreman would start out the week cutting his highest grades of lumber in the thickest sizes. Monday might be all 8/4 FAS, Tuesday might be 6/4 #1 common, and so on, until on Thursday and Friday he got around to the 4/4 upholstery grade wood. He achieved a high efficiency, but it forced the machine and sanding departments to use their people ineffectively. As a result the plant was less able to meet customer demands.

The supervisor, conscientiously trying to make his department run as "efficiently" as possible, hurt the overall productivity and profitability of the plant.

Standard Cost

Let's take another example that looks at standard cost.

Suppose you have two old boring machines. Each takes about 20 minutes to set up, and each can produce about 50 pieces per hour when running. Both machines can run the same parts, and both require a crew of two.

Now, assume I find a new CNC machine on the market that can run the same parts with a 20-minute setup and a production rate of 75 parts per hour. The machine will cost $100,000 and I will save $50,000 in labor and overhead, for a 50 percent R.O.I. The project is a go, right?

A lot of equipment has been bought based on numbers like the ones above. Although it is true that we have reduced our labor costs per part in the above example, it is also true that we have reduced the plant's overall production capacity. Look at a direct comparison of the two alternatives.

Say we begin setting up all three machines at the exact same time on the exact same part. After 20 minutes, all three machines begin to work. Each load is 100 pieces.

The new machine finishes its 100 pieces in 80 minutes. Combined with the 20 minute setup time, that gives the new machine a total time of 100 min. for 100 pieces (1 minute per piece). The old machines each take 120 minutes to finish their 100-piece load. The total time for each machine is 140 minutes. Since each machine produced 100 parts, the total time is 140 minutes for 200 parts, or (0.70 minutes per piece).

In this case, replacing the two older machines with one new CNC machining center would undeniably reduce the man-minutes per part — but it would also reduce this area's total capacity by 30 percent! If this operation controls the flow of these parts to assembly, we may have just reduced our plant's ability to ship product in order to "save money."

Utilization

Even if a capacity study shows that we have enough capacity on the new machine to meet demand, we may still find ourselves missing more ship dates.

One of the things that engineers do when trying to justify new equipment is route every part they can across the new machine to ensure it has a high "utilization." If the head office spends X thousand dollars for a new machine, they are not going to be happy to see it idle. A long line of parts starts to appear queued behind the "super-machine." To achieve high efficiency, workers will usually group like parts together to minimize changeover time.

For example, if they are running long parts, and the machine has a manual length adjustment, they will start with the longest parts and work their way down to the shortest. (This is a variation on what we saw the rough mill manager do in the earlier example.) The problem, of course, is that your plant doesn't ship parts — it ships assembled furniture.

To make a given product, you need a long part, say part Q1, that is on the machine right now. But you also need part R7, a short part, and the R7s are WAY back at the end of the line. You are going to have to wait until that machine has finished its entire queue of stock before you can build that unit. Machine utilization is high, but the production of salable units suffers.

For years managers were told to take care of their departments — make sure everything in their area ran efficiently — and the company would do well. Today we are asking our managers to grasp a very difficult fact: Ideas and principles that meant success 15 or even 10 years ago will spell the demise of their company if they aren't changed.

Managers must discard old measurements and practices that many have followed with success for decades. This is no small undertaking, but if managers don't shake off their old manufacturing mind-set and develop new ways to measure, success in today's new manufacturing environment may be unattainable.

Chapter 5

Business—What's the Point?

The next morning, Frank was at his desk. He was supposed to be getting his paperwork out of the way so he could spend some time with Wes Quinten this morning. In fact, he found himself standing by the window, watching the workers file through the various doors that lined the parking lot. Why were they here? Why was he here?

He shook his head. Geez! This was starting to sound too much like a philosophy class. His people were here to earn money so they could buy food, clothes, a home, satellite dishes, bass boats, and whatever else that made their lives comfortable and gave them pleasure. They worked here because they had to work somewhere, and their place happened to be here.

And what about Frank Average? Why was he here? He wasn't quite so cynical about himself. Yes, he worked here because he was paid, and paid pretty well. He had had opportunities here that he might never have received elsewhere. Was he here to uphold the family name or to satisfy some vague sense of guilt and responsibility for failing to keep the company afloat as an independent business?

Honestly, Frank didn't know. But he knew that he had to do something. He knew that work helped give him a sense of purpose and accomplishment, despite all of the troubles. So why not work here, in the company that still bore his and his father's name?

His reverie was interrupted by a knock on the door. Carl Chzenski, his new packing foreman, stood there looking uncomfortable. "Frank, do you have a minute?"

"Sure Carl, come on in. What's up?"

Carl sat down and clasped his hands in front of him. "Frank, I've done a lot of thinking over the last couple of days. I think I should step down as packing and shipping manager."

Frank was stunned. All of the reports from people on the floor and the older supervisors suggested that Carl was doing a good job, considering his inexperience. "Why do you feel that way? Have you gotten a better offer from somewhere?"

It was Carl's turn to be surprised. "No, nothing like that. I am just not sure that I am qualified to be a supervisor yet. I feel like I don't know what I'm doing!"

Carl summarized his conversations with Al, Benny, and Donnie. He was missing some basic concept of supervision, he said. He wasn't helping the company in the ways he should. "I've always felt that if I couldn't do something well, I should stand aside and make room for someone who can do the job. That's why I came here."

Frank smiled. He had tried to do something very similar when it had become apparent he would not be able to run Average Furniture alone. He now found himself saying much the same thing his father had said. "I am not so sure it is anything lacking in you, Carl. The furniture business — hell, all business — has gotten a lot tougher in the last few years.

"It's a credit to you that you are honest enough to admit you need help in being the kind of manager you want to

29

be. So let's concentrate on getting you that help. Who knows? The fellow who would replace you might have the same difficulty but not be objective enough to see it."

Frank stood up and extended his hand. "Let's both give some thought to how we can solve your problem and get together again later in the week. By the way, everyone else, and I mean everyone, thinks you are doing a really good job."

Carl smiled and nodded. On his way out, he passed Wes Quinten coming in. True to his word, Wes was wearing a polo shirt and khakis, which somehow managed to evoke yesterday's warm-up suit. Some folks are like that, Frank reasoned; they just don't look natural in suits. Wes shook Frank's hand. "Good morning, Frank! Sleep well?"

Frank laughed. "No, I didn't, thanks to you. I tossed and turned all night trying to come up with formulas. I was even dreaming about them."

"And did you find the formulas?"

"No."

"Good."

"Why good?"

"Because if you had, we would have had to spend time discussing why your formulas were wrong. Frank, it was a trick question. The reason you couldn't write the formulas is that the relationships don't exist. At best, you can make a general statement that profitable companies seem to measure well in efficiency and utilization. That does not mean that the opposite is true.

"As for standard costs . . . well, I'm getting ahead of myself. I want to take a tour through the plant with you. Frank, how good are your assistant supervisors?"

"In general, they are a good bunch. Why?"

"Would you feel comfortable with them running the plant in the afternoons for a while?"

Frank thought for a moment. "Yes, I think they could handle it. Again, why?"

Wes opened his thick captain's briefcase. Inside was an assortment of books, most of which Frank had never seen before. He looked at the authors' names: Goldratt, Schonberger, Hall, Ohno, several others. "Frank, my boy, you and the other supervisors are going back to school."

Frank rolled his eyes, and Wes laughed. "I don't think it will be that bad. In fact, you and your people will be doing most of the talking. Speaking of which, I need you to be thinking about one other thing. We'll need to find someone in the plant to be a champion, to spearhead the change process. It should be someone who isn't happy with the status quo, who is open to new ideas, and who isn't afraid of conflict."

Frank thought of his conversation with Carl. "I think I have just the person you need," he said, as they headed toward the plant.

The Heart of Business

That afternoon, the management team gathered at a classroom in the local community college. It was summer, and most of the rooms were empty. Wes had insisted that they go off-site to avoid interruptions. Frank had left word with his secretary that, short of a fire or accident, they were not to be disturbed.

He looked around at the dozen or so assorted people, including all of his supervisors. He wished he had been able to give them a little more warning; the whole bunch looked uncomfortable as heck. Despite his assurances, they were probably expecting some disastrous news.

Wes came in, rolling a dolly piled high with cardboard boxes. He lined up the boxes on a table near the door. Frank got up to help Wes open them. They contained copies of the same books Wes had been toting in his brief-case.

Wes walked to the front of the room and greeted everyone. "First, I want you to understand that I am not here to announce any layoffs, firings, or anything of that nature." The room visibly relaxed. "I am here because the corporation considers this plant and, in particular, its management team, to be the best one suited to try a new pilot program."

Although no one spoke, you could almost hear a collective groan. "This is not another 'Solution-of-the-Month' type program," said Wes, reading their faces. "But the only way to prove that to you is to do it. So let's get started. I want to do a little brainstorming first."

He walked to the marker board. "One at a time, and as quickly as you can, name all the purposes of this company. Why does Average Furniture company exist?

"You," Wes said, glancing at Benny's name tag. "Why is the company here, Benny?"

"To feed me," yelled Benny, rubbing his pot belly.

Wes duly wrote "To feed Benny" while the laughter crested and died down. "That is certainly a valid answer. Next?"

After about 15 minutes the list on the board looked like this:

To feed Benny	To make furniture
To provide jobs	To make money
To be efficient	To make quality furniture
To keep costs low	Make customers happy
To help the community	Get good lumber yield
Good work environment	Use materials wisely
Keep me off unemployment	Utilize resources well
Make a difference in the economy	

Wes put down the marker. "Okay, which one item up here is the most important for the company, in your mind? Any votes?"

"Of course, feeding Benny," yelled Benny again.

Wes chuckled. "I have no doubt that is the most important goal for you, personally. But what about from the company's point of view? What is the most important thing?"

Adele, the rough mill manager, raised her hand. "Well, if we don't use our materials and people well, we won't be in business long, so I would say efficiency and yield are most important."

Wes caught Frank's eye and grinned. "Okay, good argument. How many votes for efficiency?" Several hands went up. Al started to put his hand up and brought it back down. Wes noticed. "Al, do you have some other thought?"

"Well, I was just thinking, the best way to get efficiency real high would be to run just one item, say a 2800 series dresser, over and over. You'd be real efficient, and the machines that run it would be utilized real well, since there'd be no setups — but you would also be out of business real soon."

"Good point, Al. Adele is absolutely right when she says we have to use our people, our machines, all our resources, effectively. The trouble with a measurement like efficiency, or yield, is that they measure if you are using something or not, but they don't measure if you are using it well."

Wes walked to the board and crossed off those two measurements. "No offense, Benny, but ...." He then crossed off "To feed Benny."

"Remember," said Wes, "We are looking for the most basic purpose for a business to exist."

In the front row, Carl had been listening thoughtfully. Now he spoke. "Survival."

"Excuse me?"

"Uh, I said, 'survival,'" repeated Carl. "When you talk about animals, the most basic instinct is to survive. To get food, water, whatever the animal needs to survive. All other instincts are secondary. At least, that's the way they tell it on the Discovery Channel."

"Great analogy, Carl," said Wes. "And what would be the equivalent of food and water to a business, anyone?"

"Money," said Frank, thinking about the conversations he had had with Wes already.

"OK," Wes said. He circled "To make money" and put several "$$$" on the board. "Would anyone care to disagree that a company's most basic need is to make money? Or put another way, can you see anything on this list that is more important?"

"More important, no," said Nina, a mill room expediter. "But are you saying that some of the other things, like quality and customer service, aren't important?"

Wes shook his head. "Of course they are important, Nina. But what we're trying to do is help you realize that some of the measurements that seemed so important, like efficiency, may not be helping you reach your most basic goal."

Wes looked at his watch. "It's getting late. Are we all in agreement that we can accept 'To make money' as the basic purpose of the Average Furniture Company?" The listeners nodded.

"Great. I am going to give you some books that talk about this same thing. Please read the chapters that I have marked." Wes smiled. "You never know, I might give a pop quiz tomorrow afternoon.

"What I want you to do is imagine you are the president

34

of Average Furniture Company. You want to know if your company is doing well or not. What are the minimum number of measurements you would need to be able to say whether or not your company is meeting your goal? They are in those books; good luck in finding them. See you tomorrow afternoon."

Simple Beginnings

That night found Wes, Frank, and Carl back at the pool hall. They had asked Carl to come along and talked to him about being the team leader for the project. Although he seemed a little intimidated, Carl had agreed to try. They were now taking turns shooting 8-ball.

Frank and Carl were playing; Wes was sitting on a bench, nursing a tall frosty mug of beer. "So Wes," said Frank, "if things like efficiency and utilization aren't good measurements, why have they been around so long?"

"Well, Frank, they were valid for a lot of years. There was a time in American industry when you could pretty much sell anything you got out the door. If you are in a company where you don't have to worry about what you make, then efficiency can be a valid measure of your company's profitability.

"The trouble was somewhere along the way, all of the measures like efficiency and utilization became synonymous with profitability. We began to believe that you couldn't be profitable without being efficient. When the marketplace changed, American industry just keep chugging along under the same old assumptions, and couldn't figure out how foreign competitors who worked under a different philosophy could beat them in the marketplace.

"We tried to blame it on everything — cultural differences, government interference, outright illegal or immoral acts on the part of competitors. While there was some truth to all of these arguments, the real problem was that the rules of the game had changed on us when we weren't looking, and we took a long time to realize that and begin to change so we could play by the new rules."

Frank missed a shot after sinking three balls, and Carl took over. Frank joined Wes on the bench. "When we came into business, we learned how business functioned from those who were already there," Wes said. "You learned it from your father. Carl learned it from guys like Benny.

"It never occurs to most of us to question the more experienced people, especially when we are talking about the basic concepts of how business works. Most of the time, we couldn't even articulate them as concepts — we just knew that 'This is how business works.'"

Frank shook his head. "It almost sounds too simple."

Wes smiled. "The idea of re-examining the rules is simple. Putting it into action will be one of the hardest things you've ever tried to do. It is going to be a fun four months, my friend."

Frank wasn't really sure he liked the way Wes said 'fun.'

Chapter 6

New Performance Measures

Some companies have tried to adopt Just-In-Time or other modern manufacturing methods with disastrous results. When I have had the opportunity to review these attempts with the participants, two common themes stand out:

1. The managers or people who had to implement and live with the change in philosophy had not been sufficiently trained to really understand it.

2. The companies tried to adopt the new philosophy while still keeping the old measurement systems.

What these companies failed to understand is that people will perform based on how their performance is measured. It does no good to talk about learning new ways to run your business if you are still concentrating on efficiencies, labor variances, and other outdated measures in your weekly or monthly review meetings. To change your people's behavior, first change how they are measured.

In our new manufacturing environment, the new measurement system should do two things:

1. It should tie the performance of everyone on the floor to the company's financial well-being.

2. It should provide guidelines for attracting and

37

keeping customers.

Working Toward the Goal

In his popular 1984 book *The Goal,* Eliyahu Goldratt pointed out that all companies have the same primary purpose: to make money. There may be other, auxiliary goals that help define the company, but for it to achieve any of these secondary goals, it has to survive — and to survive, it must make a profit. Survival that's not based on profit defines a nonprofit organization, not a business.

Goldratt further proposed that only three measurements — net profit, return on investment, and cash flow — are needed to define how a company is doing financially. Goldratt's concepts have come to be known under the heading of Synchronous Manufacturing, a series of principles that will figure prominently throughout this book.

This all sounds logical. But can you imagine asking the person working in the wipe stain booth how much he or she was improving the company cash flow today? Goldratt realized that businesses needed to develop measurements that were easily understood by people in production and directly related to the company's financial health. The measurements he developed were as follows:

1. Throughput: the rate at which the company generates money through sales.

2. Inventory: the money the company has invested in purchasing materials it intends to sell.

3. Operating Expense: the money the company spends to turn Inventory into Throughput.

Put another way, throughput is the money flowing into the company from customers, inventory is the money currently invested inside the company, and operating expense is the money flowing out of the company.

In the early years of this country, many craftsmen ran their businesses using the "cigar box" method. The blacksmith kept a cigar box under his work bench. When someone paid him, he put the money in the cigar box. When he had to buy something, he took some out.

At the end of each week, he would count the money in the box. If there was more at the end of the week than there was at the beginning, he had made a profit. If not, he had a loss. If he had to buy something, say a new forge, he would weigh the cost of that item against the money it would bring in. If the money started going out at a much quicker rate than it was coming in, he took steps to cut his expenses and attract new business. The system we are discussing here is simply a modern adaptation of that ancient method of doing business.

These measurements, which every floor person can understand, are tied directly to the financial measurements of net profit, return on investment, and cash flow. Keep in mind that each measurement talks in terms of money and sales, not production. If something is produced but not sold, it does not contribute to the goal.

Our production goal, then, is to drive throughput up, while driving inventory and operating expense down.

Using the New Measures

Every experienced plant manager has had to "build to inventory" at some time. During slow sales periods, a plant will schedule and manufacture units that it hopes to clear out when sales rise. On the face of it, this seems like a reasonable attempt to even out the production of the plant.

Let's look at such a decision in light of the three new measurements. Since the items are going into inventory, but we are doing it to maintain the plant's previous production level, throughput would be reduced. Inventory, obviously, would go up. Operating expense would also go up, since we have to pay the order processing and carrying costs.

In this example throughput went down, inventory went up, and operating expense went up. We are moving in the exact opposite direction we wish to in terms of the goal.

So what should our answer be? One answer might be to use the lull in production to drive your backlog down. Say your production backlog is eight weeks, and that is what you are quoting customers. You might use the lull period to work ahead on your schedule and drive your true manufacturing lead time to six weeks while keeping promised delivery dates at eight weeks. This has the same net effect as an inventory buffer, because it gives you some additional response time when sales rise.

These three operational measurements can be used to evaluate possible actions and to act as guidelines for making many manufacturing decisions.

Customer Measurements

While the above measurements are a good starting point for any manufacturing company, we also need to develop measurements that cover the ways we intend to keep and attract customers. Unlike the financial measurements above, the measurements that pertain to customer relations have to be individualized for each company. There are many ways in which you can make your company stand out against the competition, including price, quality, delivery, service, convenience, selection, custom options, and design.

Most marketing gurus talk about developing your area of "Strategic Excellence." They mean find the things you are known for and good at, and improve the heck out of them so you can leverage them into sales.

Even though you may be known for one or two key facets of customer gratification, your company must still be competitive in the other areas. Dismal performance in one customer factor can nullify any advantages you have in other areas. It doesn't matter how good your quality is if your lead times are twice as long as your competitors'. A good price won't help if your product falls apart before the customer gets it home.

For most companies, the measurements that will mean the most when it comes to satisfying customers are quality, delivery, and service. You could measure your performance in these areas as shown below:

Customer Performance Factor	Some Possible Measurements
Quality	✔ Dollar amount of returns, repairs, and credits as a percentage of sales ✔ Number of customer complaints as a percentage of orders
Delivery	✔ Average order leadtime ✔ Percentage of orders shipped when promised
Service	✔ Average total time from receipt of a complaint to its resolution ✔ Average total time for customer request to a response

A Word About Price

In today's competitive business environment, it seems that every job comes down to price. However, trying to compete on price alone is a losing proposition. As someone once said, "There will always be someone less intelligent than you who will be lower because he does not know his true costs. Or there will be someone smarter than you who is lower because she's figured out a better way to do it." Certainly you have to be price-competitive, but don't let that be the sole reason a customer buys from you.

When you wish to change how your company performs, the first and most important thing that must change is your measurements of success. All the new management techniques, philosophies, and theories in the world cannot improve your company if you haven't clearly defined what you wish to achieve.

Chapter 7

How Do You Measure Success?

The same group gathered the next afternoon at the college. From their red eyes, Frank guessed that most of them had indeed read their assignments. Wes had spent the biggest part of the morning going through the company sales records, roaming around on the floor, and talking with supervisors and floor people. Frank had hardly spoken with him all day.

Despite the heat of the day, Wes stood up front cradling a cup of coffee. Frank had noticed that he was a coffee hound; that cup seemed to be attached to his hand.

Wes took a sip. "Okay, let's get started. I'll assume that everyone read last night's assignments, so we won't have the pop quiz I threatened you with. Does everyone understand the three measurements of throughput, operating expense, and inventory? Do you see how these can relate to the company?"

Everyone nodded, but Ray, the finishing room supervisor, raised his hand. "Yeah, Mr. Quentin, I understand it and all, but I still had some trouble relating it to what I'm doing today. Sure, I could see how you could look back on something you did and tell how it affected those measures, but I wasn't real comfortable with how I can gauge what I am doing today against them."

"Good, Ray! I'm glad to see you didn't just blindly accept what the books were saying. I want you folks to question anything you don't feel comfortable with — it's a good habit to develop. I've had the same problem with the measurements over the years, so this is how I define them using floor measures." He put a transparency up on the overhead:

Throughput - Due date performance by item.

Inventory - Inventory or lead time by item.

Operating Expense - Scrap, rework, and overtime by line item.

Units per hour on an item-by-item basis.

Frank perked up. "Units per hour? Isn't that the same thing as measuring efficiency?"

Wes shook his head. "The important term here is on an item-by-item basis. You see, it has been my experience that managers still need a gauge to determine how well their people perform operations. The trick is to measure that without it turning into a measure of how busy they are.

"Let's look at an example to illustrate the difference. Say that an operator has only one load show up at this station today. That load should take him two hours to do, and he does it in two hours. If he is measured on efficiency, and he works an eight-hour day, his efficiency is going to be two hours/eight hours, or 25 percent. But his effectiveness rating is two hours/two hours, or 100 percent. You see, I don't penalize operators for having nothing to do, because I don't want them to make parts for the sake of looking busy."

Frank nodded, and Wes continued. "The other measures shown are the items that the floor operators have direct control over, so they can see how they affect production.

As a plant, however, we want to measure how well the plant does overall, so we track these same measures for Average Furniture Company as a whole.

"Right now I want to talk a little more about measures we've used in the past and how they could be helping us or hurting us. Let's start by listing all the measurements that are used to tell you how you are doing in the company today. Shout them out."

For the next several minutes Wes was kept busy writing on the white board. In the end he had the following list:

Efficiency
Machine Utilization
$ Production per day
Overhead/Absorption
Overtime
Budget Conformance
Labor Variances
Scrap

"Okay, a good list." Wes now turned to a flip chart in the corner. "Now, I want you to put your customer hats on. You all buy things. What do you look for when you go buy a product?"

Again, a list was produced. Several times the consultant had to request that they slow down so he could catch up. Frank smiled to himself. "At least they are interested," he thought.

Now a second list appeared on the flip chart:

Price
Quality
Service
Delivery
Features

45

"Another good list," Wes said. "Would this list apply to just about anything you buy?" Most of the attendees nodded. "Okay, do you think that this list would be true for the customers of just about any product?" Again, most of the room agreed. "How about the customers for your products?" The nod was a little less sure now. Everyone in the room had the distinct feeling they were being set up for something.

"Now I have a question for you, and this will be your homework assignment for tomorrow. Tell me what one list has to do with the other list. In other words, tell me how the items we measure ourselves by in the plant relate to what our customers measure us by. That's all for today; see you tomorrow."

Chapter 8

What Should Be Changed

In the old game of manufacturing, the emphasis was on efficiency, utilization, and cost containment. In our new manufacturing paradigm, we must stop thinking of manufacturing as an efficiency machine and instead think of it as a profit machine. We do this by focusing on our three measurements: throughput, operating expense, and inventory.

Our focus should be to make changes that force throughput up, operating expense down, and inventory down. All we need now is to know three things:

1. What to change
2. What to change it to
3. How to cause the changes

Traditional Improvement Methods

For most engineers and managers, the problem has never been one of being unable to find improvements that can be made. Indeed, any engineer would only have to take a survey of what his co-workers think is wrong with their plant, and it would generate a list that could keep him or her busy until retirement. Usually it's a question of what changes would make the greatest positive impact on

47

a company. In most companies, that choice is usually dictated by a combination of factors:

1. The company philosophy and corporate strategies.
2. The focus of the plant management (i.e., pet projects).
3. Technological developments.
4. Environmental and regulatory changes.
5. Product design changes.

The bottom line is that most company improvement projects are launched in reaction to factors like the ones listed above. They may improve the company, they may be good projects, but there is no overriding philosophy that guides the improvement process.

Just-In-Time and TQC

New manufacturing philosophies such as Just-In-Time and Total Quality Control can cause a company to develop a new way of thinking.

In its simplest terms, Just-In-Time emphasizes the elimination of waste in the manufacturing process. That means all forms of waste, everywhere. It provides a wealth of techniques for eliminating waste — particularly in the areas of work-in-process and setup reduction. But J-I-T does not identify where to start this process; it promotes the improvement of every facet of manufacturing everywhere.

Total Quality Control, known as TQC (or as TQM, for Total Quality Management), is similar in that it emphasizes improving the quality of the entire company. Not just in product quality, but in the office, customer service, purchasing . . . virtually every department in the organization. It holds that any process can be improved and provides numerous guidelines for improvement.

Many companies have attempted implementation of these concepts with notable successes and failures. The problem again is that the philosophies stress the need to improve everything, everywhere. Most managers and supervisors, faced with trying to implement J-I-T or TQC throughout their plant, feel just as overwhelmed as they do with their present problems.

The Theory of Constraints

In *The Goal*, Eliyahu Goldratt explains that in any manufacturing operation, there are a few key factors that control the performance of the plant. Most of us are familiar with the concept of bottlenecks. On an assembly line, for example, a bottleneck is the operation with the longest cycle time. Engineers attempting to balance the line will allocate the people and tasks so that all the other operations on the line come as close as possible to the bottleneck operation in terms of cycle time.

Every plant has its limiting resources, called bottlenecks or constraints. Goldratt showed the profound effect that bottlenecks have on a production operation.

To start with, we should define a bottleneck in light of our goal of making money. The maximum money we can make at any given moment is determined by our sales levels.

If we could somehow magically produce and ship all of the products required in one second, our maximum profit equals the total backlog of sales at that instant. On a day-to-day basis, our long term profitability is controlled by our ability to ship consistently at a level equal to customer demand. Therefore, looking at our plant as a whole, we could classify all of our resources into two categories:

1. A bottleneck resource is a resource whose capacity is less than customer demand.

2. A nonbottleneck resource is a resource whose capacity is greater than customer demand.

Every plant has resource(s) that have the lowest production levels, just as every assembly line has to have one or more operations with the highest cycle time. What most of us fail to understand, however, is the profound way in which bottleneck resources affect the production of the entire plant.

How Bottlenecks Control Throughput

In most plants, bottlenecks and nonbottlenecks are intermixed in a confusing array of combinations. All of these combinations can be illustrated with four simple examples.

Say that we have two resources: resource B, a bottleneck resource, and resource NB, a nonbottleneck resource. Let's look at how these two resources can be combined in a plant.

Situation 1: Bottleneck feeding a nonbottleneck.

Bottleneck feeding a nonbottleneck

By definition, the bottleneck will have less capacity than the nonbottleneck, so the nonbottleneck would constantly be starved for production. There would be little work-in-process between the machines, since the nonbottleneck would be pulling parts away.

Situation 2: Nonbottleneck feeding a bottleneck.

Nonbottleneck feeding a bottleneck

Here the nonbottleneck is free to run as fast as possible. However, since by definition it has excess capacity, running the nonbottleneck full blast would only result in work-in-process piling up behind the bottleneck, since we know it will run more slowly.

Situation 3: Both a bottleneck and a nonbottleneck feeding an assembly operation.

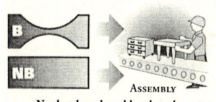

Nonbottleneck and bottleneck
feeding an assembly operation

In this combination, both the nonbottleneck and the bottleneck are free to run as fast as possible. Again, we know that the nonbottleneck can run faster than the bottleneck. If we run the nonbottleneck at full output, we will have excess parts piling up in front of assembly waiting on bottleneck parts so the assembly process can continue.

51

Situation 4: Both a bottleneck and a nonbottleneck feeding finished goods.

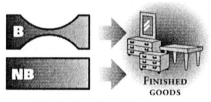

Nonbottleneck and bottleneck
feeding finished goods

In the final example, the bottleneck and the nonbottleneck resources both can run a complete finished unit ready for shipping. Once again, if we run the nonbottleneck at full capacity, we will have unwanted inventory — this time as finished goods. On the other hand, we will always be short of the bottleneck units.

In an actual plant, the "B" resource would be a single bottleneck with nonbottleneck resources before it and after it. Likewise the nonbottleneck would be a series of nonbottleneck operations. Still the basic relationships shown above would hold true.

One fact should be clear from these four scenarios: nonbottleneck resources can never be run productively at full capacity. Their level of production is always controlled by another factor — either a bottleneck resource or sales demand. Producing parts above this needed level of production only results in the creation of excess work-in-process.

In many production plants, there are no true bottlenecks; every resource is capable of producing above the level of sales. In these plants, the "bottlenecks" should really be called "pacing resources," because they control the pace of production but do not limit throughput.

So we now know what to change. We need to identify our bottleneck, or pacing resources, and look for ways to improve them.

Chapter 9

An Unexpected Bottleneck

Frank and Wes were walking through the plant. Frank was anxious to get going on the project. Although he still wasn't convinced that everything Wes said was correct, he felt that this approach was different enough to merit a try. This morning, they were supposed to begin searching for their bottleneck.

Frank was surprised when they passed the paint department and machine mill and kept walking. He was sure that the bottleneck was in one of those areas. Wes had said he had a pretty good idea where their bottleneck was. If so, it wasn't where Frank had guessed.

They kept walking until they reached the shipping dock. Wes stopped. "Here is your bottleneck, Frank."

"I don't understand. We never have any problem getting the product shipped, even at the end of the month when we flood the dock with orders. How could this be our bottleneck?"

Wes smiled. "Let me explain. You run pretty much a one-shift operation. Yes, you have to run overtime here and there, but you pretty much keep your backlog at the same level with 40 hours of plant time a week, correct?"

"Yeah, that's true. Of course, the eight-week backlog is one of our problems, but we have managed to keep it from getting any bigger than that. And we've done it without a lot of overtime. So what's your point?"

Wes began walking down the aisles in the warehouse. Mountains of cartons bearing the AFC logo stood on either side. "If you really had a bottleneck, a true bottleneck, Frank, you would be running three shifts on it, seven days a week. It would so dominate your production that you would have been able to point to it the first day I got here.

"In reality, most factories in the U.S. don't have a true bottleneck. Bottlenecks have such a devastating impact on production that, where they do exist, most managers intuitively break them. That's why most companies have a lot more capacity than they need to meet sales."

Frank said, "Well, I don't know about a lot more than needed. Sure, I have some excess here and there, but we have to scramble to keep our backlog from growing, so I don't think I have a lot of excess capacity."

Wes stopped and held up his arms, taking in the stacks of cartons all around them. "Look around you, Frank. Would you say you have excess inventory?"

Frank stared at the mass of furniture around him. "I would have to say yes."

"Would you say you have a lot of excess inventory?"

"Again, I would have to say yes."

Wes looked him straight in the eye. "It's important that you understand something, Frank. You couldn't have produced excess inventory without excess capacity. The capacity is there, because you used it to produce all this."

"So you're saying I don't have a constraint? I thought you said every plant has a constraint."

Wes raised a finger. "Ah, but I said you don't have a constraint in your plant. Your constraint is outside your plant — it's the market demand for your product."

"But if it's outside my plant, how can I use it to control

the flow of products like you talked about? In *The Goal*, they were dealing with a real bottleneck in their plant. I don't understand how you can control something that is out of your control."

Wes nodded. "Yes, it does make the approach a little different, but we can do it! And don't assume that we can't affect the market demand by what we do here in the plant, either. Remember that list of factors we talked about? The customer measurements? They were quality, price, delivery, features, and service. How many of those measurements are controlled or at least heavily influenced by what you do here in the plant?"

"All of them, I guess," said Frank.

"Darn right," said Wes, slapping Frank on the back. "But we'll get to that later. For right now, let's go back to your office. I want to show you how we use this fact to control your plant's flow."

Chapter 10
A New Approach to Manufacturing

We've already looked at classifying the resources in your plant as either bottlenecks (or constraints) and nonbottlenecks (or nonconstraints). Two key points were:

1. Where constraints exist, they control the rate of production for the entire plant.

2. Non-constraints can never be effectively utilized at full capacity.

It should be obvious by now that production improvement efforts directed at non-constraints are misplaced. These efforts only increase the excess capacity of that resource.

In *The Goal*, Eli Goldratt gives a straightforward five-step approach for improving the throughput of your plant while reducing operating expense and inventory. Let's take a look at this process and how it works.

The Five-Step Improvement Process
The five steps can be paraphrased as follows:

1. Find your plant's constraints.
2. Maximize your constraint's production.
3. Tie all nonconstraint resources to the constraint.

4. Elevate the constraints.

5. Re-evaluate your whole production system.

Step 1: Find Your Plant's Constraints

How do you go about identifying your plant's constraints? Well, if your company has detailed and accurate operation times, the process is straightforward.

In most companies, a few items or models make up a large percentage of the sales volume (this is known as the 80/20 rule). Examine each part of your big sellers and find the operations with the highest cycle time. You should soon see a pattern with a few machines or work centers showing up again and again. These are likely candidates for constraint resources.

Suppose you don't have a computerized routing system, or your data is totally unreliable. Is it impossible for you to identify bottlenecks? No. Bottleneck operations exhibit certain characteristics that set them apart from other resources:

A. Look for places where there's lots of inventory behind the resource, very little after it. Most plants try to keep all their workers and machines busy; it is a strongly rooted manufacturing habit. We know the constraint is the slowest resource in the chain. Therefore the machines behind it will be flooding it with parts, putting a lot of inventory in the queue. On the other side of the constraint, the machine it feeds will be starved for parts and will be taking loads away as soon as they become available.

B. Identify the well-known "problem spots" in your plant. The people on the floor know where orders tend to bog down. Those work centers are likely candidates for constraints.

C. Verify your hunches by tracking production through the constraint and comparing it with shipments. We have said that constraints control the production of your plant. If that is the case, then you should be able to identify a relationship between the number of finished units' worth of parts passing through your constraint and the number of plant shipments. Plant shipments should go up and down as production on the constraint moves, after allowing for a time difference from the constraint to shipping.

What if you end up having to guess at whether this machine A or that machine B is truly a constraint? What should you do first? Just pick one! The worst that can happen is that you will spend time working on a nonconstraint and creating excess capacity, and you were doing that under the old system anyway. Do your analysis, choose your most likely candidates, and get going.

Step 2: Maximize Your Constraint Production

Once you have identified your constraints, work to maximize their production time. You can take several steps to improve throughput quickly:

A. Stagger lunches and breaks so that the constraint continues producing during these times.

B. Are you running three shifts on the constraint? If not, consider it. You might have to install a stand-alone dust collection system or compressor for this machine, but if it increases plant production, the payback will come quickly. Downtime on a constraint should be limited to preventive maintenance.

C. Inspect parts before running them through the constraint. (Time spent processing parts that are already bad means fewer units going out the door.)

D. Identify parts that have gone through a bottleneck so that operators will take extra care with them.

E. Off-load parts that can be machined on nonbottlenecks, even if the nonbottleneck operation takes longer or costs more to produce these parts than the bottleneck machine. The increase in revenues from shipping more product per day should more than offset the "increase" in labor, even if you have to subcontract the operation.

Step 3: Tie All Nonconstraint Resources to the Constraint

This one can get involved, but the concept is fairly simple. We have said that a nonconstraint can never be run at full capacity. So what level should a nonconstraint run at? Its level of production should be roughly equal to the level of production for its constraint (either a bottleneck or the market demand). Setting your nonconstraint production at this level will help to reduce inventory — one of our three main goals — without hurting either throughput or operating expense.

For example, if a nonconstraint machine could produce 200 units a week at full production, but a constraint downstream will only allow you to produce 100 units a week, there is no need to run more than 100 units a week through the first machine. Those additional parts will only pile up behind the bottleneck.

To fully implement this step, you must tie the release of materials to the floor to the bottleneck production rate. (The methods for doing this will be discussed in the chapter on material flow.)

There is one unavoidable result of running machines at less than full capacity. People on the floor are going to have idle time. This is probably the single most difficult concept for managers, supervisors, and even floor people to deal with. In the business world, it's one of the most basic ingrained values: keep your people producing.

The idea that we would force them to not work on parts seems downright un-American. You need to develop lists of alternative activities that operators can do during

production lulls. Two obvious tasks are cleaning and preventive maintenance. Other items might be problem
solving or helping to work on constraint operations.

Step 4: Elevate the Constraint

Now we get to the task of trying to make the constraint
produce more products. This is where a lot of the
Japanese J-I-T tools and traditional work method techniques can be very helpful. There are many books on
these subjects.

In summary, the best ways to increase a constraint's production are:

A. Reduce changeover time. Techniques that can be
used include:

• The SMED (single minute exchange of die) system
developed by Shigeo Shingo at Toyota.

• The ORE systems developed by Lee Houston at North
Carolina State University. (The letters stand for organize,
remove, and eliminate.)

• The WERC (or woodworking equipment reduction of
changeovers) systems. Essentially, this is is an adaptation
of Shingo's system to the woodworking industry I developed a few years ago. (For a discussion of the WERC
system, see Appendix A.)

To reduce changeover times, you go through your setup
operation, preferably filming the entire sequence, and
break the steps down. You then determine what operations have to be done while the machine is down, and
make sure all other steps are done while the machine is
running. You may use techniques such as having multiple
people work on the changeover (like a NASCAR pit crew).

You also can modify the machine, tools, fixtures, or

other elements to make them easier to locate and attach.

B. Standardize work methods to improve consistency. There are few operations that can't be improved by standardizing methods, layout, and tools. There is a mountain of textbooks on these techniques, and the methods are similar to those used for changeover reduction.

Step 5: Re-Evaluate Your Whole Production System

As you begin to implement changes, you will be amazed at how quickly you become attached to new methods of production, scheduling, and running your operation. Be careful! You need to re-evaluate the entire plant again at step No. 5.

For example, if you have been successful in breaking a bottleneck, but still have the plant's production tied to those former constraints, it could be preventing you from improving even more. The main idea here is, don't take anything for granted when you re-evaluate your plant. Look at the plant as though you were starting for the first time.

No In-Plant Constraint

All this sounds great in theory. But suppose that your investigations indicate there is no real bottleneck in your plant? In fact, unless you run a three-shift operation on at least some of your machines, it's very unlikely that you have a true bottleneck. Most plants in the U.S. don't have true bottlenecks; their constraint is the market demand for their products. If that is the case for your plant, how do you go about following the guidelines?

Obviously, it's more difficult to elevate the constraint (increase market demand for your products) in this case, since you don't have direct control over the process. It's not impossible, however.

One step you can take right away is to reduce your inventory and operating expense by tying them to your market demand. For example, let's say that you have an item, X, that sells at the rate of 20 units a week. You can extend that back to how much lumber, particleboard, and other materials you need to bring in for that unit.

If you do that for each unit you produce, you get an idea of how much raw material you are going to use every week, and how many units you are going to ship out each week.

You then set up a pull system to pull products through the plant as they are used. The quantities used to set up the pull system are based on two factors only: the average usage of the item and the time required to replenish that item. For example, if you use 20 units a week of an item and it takes two weeks to replenish the materials, then the pull quantity should be set at 40 units per week. Using this system will allow you to set quantities that minimize inventory while ensuring order completion.

Flow Materials In and Out at the Rate of Sales

40 units every 2 weeks
or 20 units per week

Stock buffer

40 units every 2 weeks
or 20 units per week

The principles of Synchronous Manufacturing are simple to explain, but it takes time and patience to apply them. Remember that you are trying to create a new way to manufacture. Accept the fact that you will make mistakes; they are a part of learning a new way to work.

Chapter 11

The Transformation Begins

Frank walked into the new office he had set up for the task force. There were a few desks, taken from obsolete inventory. "No shortage of those, unfortunately," thought Frank. A couple of PCs, a white board, a flip chart. It hardly looked like the kind of place where a revolution would begin. But that was just what Frank hoped would happen here. He still had his doubts, but maybe, just maybe, this Wes guy really had something.

They had completed the informational meetings the day before. Now, at least, all the front line supervisors would know what was coming. Later they would have shorter sessions with every person in the plant, as their area was affected.

After their meeting, they had asked for volunteers to serve on the task force. Frank had stressed that this was not a temporary assignment. It would mean a permanent relocation to the team. He had been surprised at how many of his people had been willing to do it. In the end, they had ended up with a good mix of people from the plant.

He had already talked to Carl, of course, and he had agreed to head up the team. The other members were:

Lottie, from the computer room; Angie, one of the two junior accountants from the plant; Kevin, an industrial engineer; and Donnie, the plant scheduler. Donnie would have to continue to help his assistant to schedule the plant during the transition, but as he had said, "Heck, Frank, it sounds like I won't have a job when you're done, so I better learn how to make this new system work."

The team came into the room at around 7:30 a.m. carrying boxes. Frank watched as they all chose a desk and placed their items. Photos, wall plaques, souvenirs, and other personal effects all took their place on the desks and shelves. These were the little pieces of their personal life that everyone brings into their workplace to help define who they are.

There wasn't much talk. With the exception of Donnie and Carl, most of the people had only a passing acquaintance with each other. Frank wondered, not for the first time, whether this diversity of backgrounds would be a help or a hindrance, but Wes had insisted on a cross section of the departments.

Wes came storming in at 8:00 a.m. carrying a box of his own. He entered with his usual booming, "Good morning!" and walked to the center of the office, where a mini-conference room had been set up. He set his box down in the center of the table and opened it. It was then that Frank noticed his shirt. It was a typical golf shirt, but over the breast pocket was the inscription "Average Flow Team" in blue. Over the "Average" had been written "Above" in a hand written red font. So they were the "Above Average Flow Team," huh? Kind of cute — but Frank decided he liked it.

Wes started handing out shirts to the team. "I want to welcome you all to the team! You have been chosen to do nothing less than transform your company. That's a big honor, and a big responsibility. There are going to be days you regretted volunteering for this team, but I can almost guarantee that at the end you will be very proud of what you have accomplished. We have an aggressive schedule, so let's get started."

Wes spent about an hour going over how they should tie the ordering of raw materials and the flow of work through the plant to sales levels. He showed how they would use pull tickets to move the materials through the plant and to signal replenishment of work-in-process. Frank raised his hand to interrupt. "Wes, I don't see the difference between your pull system and what the Japanese call a Kanban system. Aren't they the same thing?"

"They are similar," Wes said. "But the differences are significant. The pull system, like the Kanban system, signals the replenishment of materials. The Kanban system, however, tries to continually force inventory levels down to the absolute minimum. The idea is that you force inventory and manning levels down to where problems occur in meeting orders. Then you solve the problems, and thereby continually improve your process. Also, in the Kanban system, there's no distinction made between constraint or pacer resources and other resources.

"With the pull system, your primary concern is ensuring throughput, so you place inventory where it will help ensure that goal. Also, we take into account the location of constraints or pacers when placing inventory buffers."

"So how do you decide where to place the buffers in your pull system?" asked Lottie.

"There are a series of guidelines to follow, depending on the type of plant you have. But in general, the rules are these:

1. Locate buffers to protect bottlenecks, if they exist.
2. Locate buffers where parts diverge or converge. In other words, where parts split into different parts or are assembled. Rather than a hard and fast rule, the individual divergent and convergent locations should be reviewed on a case by case basis. We call these locations central points."

"You might have to go over that one again," said Angie. "I'm afraid I don't know much about how things work on the floor."

"No problem, Angie." Wes walked over to the white board and began drawing a diagram. "Most furniture plants make a large variety of end items from relatively few raw materials."

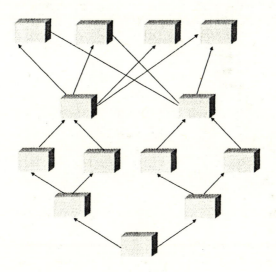

"Man, it looks complicated," said Carl.

"Sure it is, and it's that very complexity that scares most people off," said Wes. "What a lot of people don't realize is that you don't have to control every operation. There are control points that can allow you to reduce inventory and improve throughput at certain key points.

"The obvious place to look for locations for buffers, aside from bottlenecks, are the places where the parts split. However, if you put a buffer at every point parts split, you would have an unmanageable number of buffers and pull locations.

"Instead, you have to look at where the control points for the operations occur. One obvious location is in front of assembly. The important thing is to look and see where it makes sense to set up the buffers. You'll feel more comfortable with that after we have installed a few product lines."

Wes looked at the team members. "Look, I'm sure each of you is a little uncertain about his or her role on this team. You may even think you don't know enough about production, or scheduling, or other operations to contribute. But each of you understands how a critical part of this company functions. Between you, you either know the answers or know where to go to find the answers to just about any question that will come up.

"Keep that in mind, and make it a point to share your expertise with the others. All right, let's get started by analyzing your product lines and putting together a game plan for the project."

Chapter 12

The Need for Purchasing Controls

The concept of material flow is used to refer to all of the functions that in the past have been grouped under purchasing, receiving, tooling, supplies, and shipping. The idea of material flow initially can be difficult to grasp because these areas haven't been grouped together traditionally.

For example, in most larger companies the purchasing department handles raw materials while the maintenance department buys the tooling. The reason we group them together here is to emphasize that all purchasing should be done to support the goal of increasing profits. It really doesn't matter whether you're buying hardware to apply to a door or a shaper head to apply a profile to a top. They're both purchased for the primary purpose of adding value to the product and helping you obtain profits.

A Balancing Act

Anyone who has ever worked in purchasing knows what a balancing act it is. You want to minimize inventories while making sure items are there when they are needed. Problems that occur in purchasing are generally parallel to the problems that plague manufacturing.

For example, the "big batch" syndrome that's prevalent

in manufacturing forces the purchasing department to buy in large quantities. This tendency is reinforced by the perceived cost savings associated with "buying in bulk."

In the past, purchasing decisions were largely dictated by a concept call the "Economic Order Quantity" or EOQ.

Another Sacred Cow

Of all the sacred cows that have been discussed so far, the Economic Order Quantity concept ranks right up there with efficiency and utilization. In theory, the concept seems sound, and it's still widely taught in business and economics classes across the country.

The EOQ for a particular item is defined as the lowest total variable cost of an item that must be carried in inventory. What accountants attempt to do is figure out how much it costs to order an item (freight, order processing, as well as other purchasing costs) and how much it costs to carry an item in inventory (storage, taxes, theft, spoilage, and so on). Then they calculate the inventory level for each part.

Because ordering costs per item tend to go down as you order more (the "bulk buying" concept again) and carrying costs tend to go up as you store more items, you should be able to arrive at a lowest total cost per item. This is the

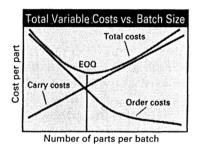

Number of parts per batch

lowest point on the EOQ curve and is the economic order quantity.

In theory, this sounds like a good way to minimize the costs associated with buying and storing inventory, but the concept is flawed in several ways.

1. There's no factor for potential lost sales if an item is out of stock.

2. There's no way to account for the potential increase in sales from reduced lead times.

3. The EOQ formula doesn't account for the major impact batch size has on lead times and throughput.

4. The EOQ formula assumes that the batch size you order must be the same as the batch size you run through the plant. But this won't always be the case.

Also, from a practical standpoint, few companies ever had good numbers to plug into an EOQ formula. They usually just calculated gross percentages which were as much guesses as good data.

The biggest problem, however, is the same problem we have had with other measurements. The EOQ formula was developed to minimize costs, not maximize profits. There's a very real difference between the two objectives.

So if we don't use the EOQ concept to figure out when we should order product and how much we should order, how do we do it? The answer lies again in the system constraints or bottlenecks.

We have said that the ability of a plant to produce product is determined by either its bottlenecks or market demand, never by nonbottlenecks. In any plant there are very few bottleneck or constraint resources. In fact, by definition each plant can only have one constraint — only one operation can take the longest time.

The rate at which the plant will process each part is dictated by this constraint. In the same way, the plant will consume the raw materials that part is made from at the same rate. If we can somehow tie our purchasing needs to our production at the bottlenecks, we should be able to meet our goals — maximize throughput with minimal inventory and operating expense.

Purchasing by Exception

The method of achieving optimal material flow in Synchronous Manufacturing is called the Drum/Buffer/Rope concept. Basically this method:

1. Determines the rate at which bottlenecks or the market consume parts — the drum beat,

2. Establishes a needed level of safety stock to protect the bottlenecks — the buffer, and

3. Ties the purchasing and release of materials to the bottlenecks (the "rope").

The D/B/R method is the best way to control material flow. But suppose you don't have a complex operation, or aren't large enough to consider implementing such a system. Then you might consider something I call "Purchasing by Exception." Purchasing by Exception (or PBE for short) uses the same basic concepts as D/B/R in a simplified form. The best way to illustrate it is with an example.

Purchasing by Exception — An Example

Suppose we want to apply PBE to the purchase of one of our items, say 6/4 particleboard. We start by tracking all the operations that 6/4 board goes through. Although there will be some exceptions, we find that the typical sequence of steps is as follows:

Operation
1. Panel Saw
2. Abrasive plane
3. Press
4. Tenoner
5. Edgeband
6. Profile sides, if it has thick wood edges
7. Profile ends, if it has thick wood edges
8. Sand, if it has a veneer face
9. Assembly

After analyzing cycle times for various 6/4 parts, we find that the press operation is the constraint on practically all of these parts. By looking back at past production, we further determine that 6/4 thick board makes up about 50 percent of all the board we process.

Based on 50 percent of total press production, we calculate that the press can run the equivalent of 100 4-foot by 8-foot sheets a day. Therefore, our average weekly consumption is about 500 sheets. Our board supplier is fairly reliable and located only 100 miles away, so we decide a safety stock of one week is sufficient. The lead time from order to delivery is four weeks.

We meet with our supplier and say, "Look, we want to partner with you on our purchase of 6/4 board. We'll give you a blanket purchase order for the next six months, but we want you to calculate a price for us based on that quantity.

"We want you to lock us in to your schedule for 500 sheets of 6/4 every week. If I don't call you by Wednesday noon, you can assume that the standard order of 500 sheets applies for the next open order. If I do call and make an adjustment, the adjustment will apply to the next open order. For example, if I call up in Week 4 and tell you to double my next order, that double order will be for Week 8."

How would this system be superior to what a purchasing manager does today? Well, consider what the purchasing manager does today. He knows his lead time is four weeks from this vendor, so if he places an order today, he knows he better have enough material already here or on the way to cover the next four weeks.

Assuming a safety stock of one week — although it

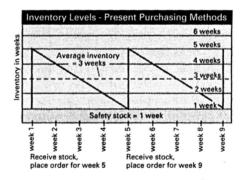

would likely be higher in this case — the company could have as much as five weeks inventory or as little as one week, depending at what point during the order cycle you were at. Assuming a steady usage of material, our average inventory would be three weeks worth.

No Worries

In our new scenario, we would have the one week of safety stock, but our high inventory level would only be two weeks. In this instance, we have an average inventory of only one week for a 50 percent reduction! It also makes it less likely that the plant will get into trouble because someone forgot to order something, or to report that an item was needed. The only time a purchasing agent needs to intervene is when something unusual happens — an exception.

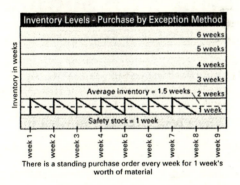

There is a standing purchase order every week for 1 week's worth of material

Okay, you say, but what if we get a huge order that we have to get out in four weeks, and it requires 2,000 sheets of 6/4, what then? My friend, you're dead anyway! Seriously, that's the entire reason for basing the material quantity on the usage at the constraint.

No matter how quickly you get the materials, you won't be able to get it through your constraint fast enough. It does no good to get all that material in and have it sit behind the bottleneck for weeks, it won't get the order out any quicker.

By now it should be obvious that tying the purchase of materials to your constraints can both reduce inventory and improve the flow of materials. Reduced inventory costs and improved responsiveness will be critical to your success in manufacturing.

Chapter 13

Working With Suppliers

Frank sat on the end of the table. This was the first meeting between the team and some of the company's key vendors. The morning had been spent with Wes giving the vendors a brief overview of what the company was trying to do and why. Now they were getting down to the meat of the discussion.

The project had been going pretty well, but not without some problems. It had seemed to take forever to get the data together and install the first pull cards. Quantities on some items had to be changed; the cards had been redone a couple of times to work better in the plant. Some plant workers had been unclear or unsure of their new roles and had been retrained. All in all, though, this first phase had gone relatively smoothly. Now the team was starting to dig in to the question of purchased parts.

As Wes came back and sat down, one of the vendors, the carton supplier, said, "Okay, I think we all understand why you are doing what you are doing, but what exactly do you want us to do?"

Wes looked at Angie, who had been chosen to be the team spokesperson for this meeting. As the project had gone on, Wes had handed off more and more of these

types of responsibilities. Angie walked up to the front of
the room and placed an overhead on the projector.

"First, we want to get the most frequent delivery
schedule we can — daily if possible. But, no B.S. here
folks, we want to know what you can deliver, not what
you are willing to promise.

"Next we want to negotiate with you to be our supplier
for a six-month period. Prices will be based on the
projected volume for that six-month period with
adjustments made at the end of each six-month period.
We'll provide you with the projected usage for each item.

"Third, we want to improve communications between
our companies. This will include visits to your companies
by our flow team and the formation of a 'rapid response
crew' to solve quality and supply problems."

The carton supplier, Irv, seemed to be acting as a
spokesman for the vendors. The team had put together a
series of objections the vendors would probably come up
with and rehearsed their responses. Irv now brought up
the first objection they had anticipated.

"If we lower the amount of items we ship to you at one
time, there's a good chance your freight costs are going to
go up. Have you considered that?"

Angie didn't miss a beat. "We have considered that, and
if it is unavoidable, we are prepared to eat the freight
costs. However, we also want to find ways to avoid those
costs. For one thing, we may ask you to share trucks with
other vendors. We may look at setting up a 'milk run' —
having one truck make daily runs to several vendors and
then deliver the items here at the plant. If necessary, we'll
even look at running trucks ourselves."

Lenny, who supplied hardware, brought up another
factor. "All this sounds good, but doesn't all this boil down
to shoving your inventory back on your vendors, on us?"

Angie smiled, they had covered this one too. "Our goal is first, to reduce our inventory. So if the only way to do that is to force you to hold it then, yes, that's what we want you to do. Think of it as your contribution in return for getting guaranteed business for six months.

"However, forcing the inventory back on you is the final solution, as far as we are concerned. Think of it, you're going to have fixed quantities to run for the next six months. For the next six months, when you come in on Monday morning, you're going to know exactly what the Average Furniture Company is going to need. If your processes permit, why not set up your plant to flow the parts through like we are doing?"

"And why would we do that?" asked Lenny.

Angie smiled again. "For the same reason we're doing it, Lenny. Your company will make more money."

The discussion continued for a few more minutes, but each of the vendors agreed to meet again in two weeks with the answers they were looking for. After that meeting the real negotiations for the six-month contracts would begin.

Outside the meeting room Wes and Frank paused to get a cup of coffee as the others headed back to the team room. They were standing there talking when one of the vendors walked by.

"So, what did you think?" asked Frank.

The vendor shrugged. "It shouldn't be a big deal. We're already doing daily shipments for several of our large customers. We were just waiting on you folks to ask for it." The vendor continued on out to the exit.

Wes and Frank looked at each other for a long moment before they started laughing.

Chapter 14

A Quick Fix for Reliability

Before you can maximize any manufacturing organization you have to understand the role quality plays in the new game of manufacturing. First, let's examine some of the results of implementing the strategies we have outlined for manufacturing and materials flow.

- We've decided to tie all of our production to the bottlenecks.
- We've also decided to tie material purchases and material releases to the floor based on the production level of our bottlenecks.

Put Idle Time to Work

In essence, we're forcing the entire plant to work at the pace of our constraints. This is good for minimizing inventory and operating expense, but it also means that people working on nonbottleneck machines are going to have idle time. This idle time makes many plant managers and supervisors nervous. What do we do with it?

The best answer is: anything but make inventory. But we need to consider ways to utilize this time. One good way to use these idle workers is to have them inspect

parts that go through bottlenecks before they are processed at the bottleneck. Because processing time on a
constraint is so important, we need to ensure it works only
on good parts. By making sure only good parts go into the
constraint, we actually increase plant throughput.

Another way to utilize this time is to have workers do
rework and salvage on bottleneck parts that have already
been processed. We want to make sure workers
understand that parts that have been run through a
bottleneck need to get extra care, but sometimes errors
will still occur downstream. If we can utilize what would
be idle time to rework these parts and use them, we have
again increased throughput.

As long as the extra activities, inspection, and rework
are done during idle time, they're essentially free. In fact,
if the manning of the plant is unchanged, and we are getting additional units out, it actually increases profitability
with no increase in inventory or operating expense. We
are meeting our goal.

Focus on Quality

But what about improving quality in the traditional
sense? Over the last 15 years or so, quality has evolved in
three major areas — Statistical Quality Control (SQC),
Quality Function Deployment (QFD), and In-Process
inspection, or "Poka-Yoke." A multitude of books has been
written on each subject. All are good tools to use to
improve quality, but in this approach we're going to concentrate on the third — In-Process Inspection. This is the
one that can be used to improve throughput directly, and
it's probably the least talked about of the three subjects.

"Poka-Yoke" is a Japanese term that means "to avoid"
(yokeru) "inadvertent errors" (poka). If I tried to write

a formula for defining a defective part, it might look something like this:

Material + Mistake = Defect

That means for a defect to exist, someone, somewhere has to make a mistake. The mistake might be in processing, specifying of materials, defining of procedures, or literally any phase of production. It could occur at the supplier, at a machinery manufacturer, at a toolmaker, or in-house. But somewhere a mistake was made to create the defect.

Poka-Yoke is used as a means not to eliminate the defect, but to eliminate the mistake that caused the defect. It is a powerful concept, and one that has played a large part in Japan's well-deserved reputation for quality products. The concept of Poka-Yoke is to build into the machining process (or a downstream process) checks and devices that help people avoid mistakes and make good products the first time. There are five major types of Poka-Yoke devices:

1. Guide pins
2. Sensors/alarms
3. Limit switches
4. Counters
5. Checklists

The best way to understand Poka-Yoke is with some examples.

Example #1

Suppose you're making the part shown here. The two

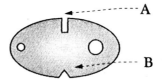

holes are bored first, then the part is notched at "A" and "B" in two separate operations.

You have a problem with the operators aligning the parts correctly when notching. Sometimes the parts come out with the notches in the wrong place or on the wrong side.

How do we go about preventing these types of errors? One way would be to install guide pins on our notching dies. The pins would be different sizes, so the part could only go on the fixture one way. Positioning errors would be eliminated.

We can actually do more than that. Let's say that the triangular notch at the bottom is made first. Create a fixture, such as the one shown below, that checks the location of the holes before making the rectangular notch.

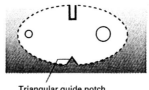

Triangular guide notch
checks the location.

Example #2

A furniture company boxed up hardware for some of its units in an automatic machine. The machine would drop the hardware in the box, seal it, and send it down a conveyor to be packed with the unit. Occasionally the hardware machine jammed, and an empty box traveled down the conveyor.

Solution: An empty box was lighter than one with hardware, so the company installed an air jet on the conveyor. The air flow was adjusted so that empty boxes would be blown off into a bin, while boxes with hardware continued on down the conveyor.

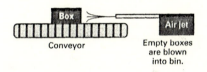

Conveyor

Empty boxes are blown into bin.

Example #3

Operators on a drill press sometimes failed to bore dowel holes deep enough. This resulted in cracked parts when the dowels were driven. To solve the problem, the plant installed a limit switch at the proper depth. If the drill was activated but the limit switch didn't trip, a

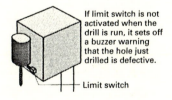

If limit switch is not activated when the drill is run, it sets off a buzzer warning that the hole just drilled is defective.

Limit switch

buzzer sounded. This ensured that the drill went to the proper depth each time.

Applications such as these are inexpensive, one-time fixes that ensure that a particular defect will not occur again. Many other examples could be shown, but the idea is to cause the operation to stop if the proper alignment, hole depth, thickness, or other characteristic isn't present.

Although it's probably the least glamorous of new quality methods, Poka-Yoke quickly can improve the reliability of any operation. When applied to a plant's constraints, it can increase the throughput of your entire plant, and its profitability.

Chapter 15

Throughput Gets a Boost

The team was taking a plant tour with Jim Pearl, the manager of quality assurance, and Frank. Wes wasn't in the plant today, so Carl was leading the team on a fact-finding mission.

The team had found that quality problems were causing several parts to turn more rapidly than they should. There was also concern that the increase in throughput the company had experienced was putting a strain on the finishing room. Frank smiled; if he had to have problems, he was glad they were because of an increase in throughput.

The team stopped first at the paint line. Frank was interested in how Carl would handle this; it was really his first test as the team leader. Carl had been enthusiastic and supportive, but a little tentative. The team stopped at the end of the line and watched the finished units coming through. Ray Wallace, the finishing room supervisor, came over.

"So we're starting to incur some overtime to meet orders here, now," said Lottie.

"It's not too bad now, a few hours during the week," said Ray. "But we're going to have to consider at least a partial second shift soon if we want to keep running as many finishes at once as we are."

"Why does the number of finishes make a difference on the finishing line?" Lottie wanted to know.

"Well, in the past we would only have one, maybe two finishes on the line at one time," said Ray. "Now we can have as many as four or five on the line.

"We used to be able to run out of our day tanks for most of the stains we used. Now, 'cause we're switching so much, we pretty much have to run out of the pressure pots. Also, we lose one to two pallets every time we change over, to give the operators time to switch their lines."

The team was watching items come off the line. "Can you run the line during lunches and breaks?" asked Lottie.

"It would be kinda difficult," said Ray. "It's not like a robotic line, or a dip system, where the parts are painted automatically. The line has to be pretty much fully manned to run."

"Are there items that you could double up on?" asked Kevin. "Run two or more per pallet?"

"We do that now, when we can," said Ray.

Kevin started to ask something else when Angie asked, "Ray, is there an empty pallet between each of those tables?"

Everyone on the team looked. Sure enough, there was a series of dining tables coming off the line. The tables were long and wide, and overhung the pallet on four sides.

Ray replied. "Well, since the tables are so big and wide, we leave an empty pallet between each, otherwise, when they go through close packed areas, they'll bang into each other."

"Ray," said Kevin, slowly, "how much production do tables account for?"

"Oh, probably about 5 percent. Why?"

Kevin took out his note pad and started sketching something. "Suppose we redesign the stands the tables set on so they're at two different heights? Then, as long as we alternate heights between each pallet, we wouldn't have to leave an empty pallet between them. That would gain us an extra 5 percent in production right there!"

"Say, that's not a bad idea," said Ray. "We'd just have to make sure the tables would ride through the system, but that shouldn't be too tough. I like it!"

As they were walking through the finishing area, Angie stopped by some booths which lined the side wall. These booths weren't connected to the main line.

"What are those for?" Angie asked.

"We use those to run some really large pieces that won't ride the line," said Ray. "Sometimes we run show samples over there."

"So you could run production through there?" asked Carl.

"Well . . . yeah, I guess you could, but it wouldn't be very efficient."

"Ray, you've been in the seminars, you should know by now that efficiency isn't the big concern," said Kevin. "If you manned the line, could you run the odd-ball colors and custom colors on it, to free up time on the line?"

"It would have to be the right people, but I guess we could look at it."

The team discussed some other ideas that might gain them some additional capacity, like subcontracting some of their finishing duties, and then walked back to the office.

"I wish all of our problems were that easy to solve," said Angie.

"Don't think we've solved it," cautioned Frank. "All we've done is buy some time. But now we can take a hard

87

look at other ways we might create more throughput. In the meantime we'll be getting about 10 percent more through a maxed-out resource. Good job, team!"

Chapter 16

Product Design in the New Game

We're going to leave the production floor to talk about how this transformed company you're working toward should handle product design. We're also going to spend some time on the definition of your product. It may surprise you to learn that this definition is different from what you think it is.

Let's assume for a moment that you're well on your way to achieving your new production goals. Lead times are dropping, inventory is dropping, throughput is up. Now here comes the marketing guys and product design engineers wanting you to introduce a new line and screw everything up.

The good news is, your leaner, more synchronized factory will be better able to handle a prototype run than your old plant was. Because there's less inventory in the factory, you'll be able to move prototype parts through faster. And, because you now have idle time (or, as we call it, opportunity time) on your nonbottleneck operations, you can run prototype units during off time and not disrupt the flow.

You'll have to be careful about how you run the prototype parts through your constraints, but at least now you know that's where you need to be careful about losing time. So, all in all, your factory is better prepared to run samples when needed.

But there's another aspect of this new competitive world we live in — product life cycles are shrinking. In residential furniture, it wasn't unusual from the mid-1970s to the mid-1980s to have a line run for 10 to 15 years. Today, you're fortunate to get a three-year life span out of a product line. Soon we'll be turning our entire product line every year.

What does all this mean to the product design process? It means that everything we do will have to be changed fundamentally to continue to compete in the future.

1. Product lead times must be shortened — drastically. If a product has a life span of one to two years, we can't take a year to design and introduce it. If we did we'd be perpetually behind.

2. Product design must take into account both the physical item being sold as well as any soft aspects of the product. (More on this point in a moment.)

3. Product design must cease being a crap shoot. We have to know that the products we're going to introduce will sell.

Shorten the Design Cycle

Product design cycle time needs to be no more than one-quarter of the expected product life cycle time. If your product's life cycle is one year, the time from conception to introduction shouldn't be more than three months. How can we produce new products in this amount of time? It will take a combination of several factors.

First, we'll use cross functional teams to eliminate hand offs between departments. The team will have representatives from design, engineering, production, purchasing, and tooling/maintenance. Suppliers also may need to be a part of this process. This will eliminate the "throw it over the wall" syndrome that has been prevalent in the past.

Next, we should design our new products around a basic structural shell. Within each new series or product line, there will be certain units that are always present. In a dining suite, for example, there would be one or two buffets, matching chinas, dining tables, chairs, and a server. We should strive to make these units as common as possible in terms of overall dimensions, interior components, and other factors.

There's no need to design each unit from scratch; just make sure the customer's requirements are met. Many companies spend a lot of time and money to go back over existing products and do standardization work; it's much easier to design it in on the front side.

We should use Quality Function Deployment (or QFD) and focus groups to determine what characteristics are important to the end user and incorporate them into our design parameters. What that mouthful means is that we should ask our customers what features are really important and make sure all of our products have them.

The QFD process was developed by the Japanese to determine the likes and dislikes of their customers and distill that information into design elements that would be included in their furniture. The same process can be used to improve the acceptance of your furniture.

For example, let's say you're designing a line of juvenile furniture, and you want to know how you should make the drawers. You select a group of potential customers, and have them look at several different types of drawer boxes. Is solid wood important, even at this price point, or would they prefer to see easy-cleaning melamine drawers? How deep and how tall should the drawers be? How many clothes drawers and how many "junk" drawers should be included in each case? By doing a number of these studies and compiling the data, you'll soon find the best combination of features, according to your customers.

More Than a Box

Woodworking manufacturers are coming to realize that their product is more than just the box they ship out the back door. It's also the service, the reliability, and other service factors associated with buying that product that make up its true identity.

This is so vital a concept that it needs to be emphasized. In the production world we're hung up on the product, it's what Tom Peters called the "lumpy object."

But customers looking to buy something ask themselves, "What benefits will I gain from giving up my money for this item?" Those benefits are more than just what the product can do. Yes, it's the product, but it's also the delivery, the service, responsiveness to customer queries, and a host of other factors.

The perfect illustration of this is a trip I once took in the field with a salesperson to visit some dealers. (A note to managers and engineers: Go out in the field and look at your products on your customers' receiving docks. It can be both an educational and a humbling experience.)

This particular customer began by saying that our company built the best products in the business. Then, for the next 90 minutes, he explained why he sold a lot more of another company's product. By and large, it was a lot of little things — faxing acknowledgments overnight, using a single reference number on each item so you didn't have to go digging through mountains of paperwork to file a claim or return an item, always using the freight company the customer preferred for shipments. Little extras like that. It was during this conversation that I realized that we made a better piece of furniture, but the competitor offered a better product.

Two Customers

We actually have two sets of customers — the dealers and the end users. Each of these customers has different requirements and expectations, and both must be satisfied. You could say that it's the design engineer's job to satisfy end users with a product and price they like, and it is the job of the sales and marketing departments to satisfy dealers with terms, conditions, and services they like.

And manufacturing? We have to satisfy both sets of demands. That's why your flexible, synchronous factory is so important.

Chapter 17

Not the Same Old Marketing

The team was meeting with Alan Zaytsef, the head of product engineering, and Joe Westbrook, vice president of marketing for InterMast. Joe was responsible for the marketing of Average Furniture Company's products. The two were here today to talk about new product introductions.

As the throughput of the plant had increased — and the potential constraint at the finishing line had been eased — it had become apparent that the plant could produce a lot more than was presently being sold. The team was here primarily to start a dialogue in an attempt to kick start some new product ideas.

Frank had been held up by a minor emergency in the rough mill and arrived at the meeting about 10 minutes late. Apparently he had missed a good beginning, because when he walked in Joe was saying, in a none-too-soft tone, "But you can't do that!"

Frank took a seat by the door. His plant manager's instincts pushed him to jump in on behalf of his people, but he had been urged by Wes to let the team handle these types of meetings. After the project was underway here, the team would be expected to travel to other InterMast

plants to help in the implementation of Flow Manufacturing. He knew they needed to learn to handle
confrontations.

Carl's confidence and leadership had grown greatly over
the last few weeks. He answered back, "Why, Joe? What is
so sacred about the furniture markets? Why do new
products have to be introduced there?"

Joe sputtered, "Because, well, because they do! Look,
the entire industry introduces its new products at the
spring and fall markets in High Point. If we introduced
items at other times, what would we show at the market?"

Carl shrugged, "The items that had been introduced
since the last market. What's wrong with that?"

Joe shook his head. "But they wouldn't be new! The
competition would have had a chance to pick them apart,
or to even knock them off if they were successful."

Surprisingly, support came from Alan Zaytsef. "Come on,
Joe. That argument doesn't hold water. You know how we
are pushed to just get out our own new products for the
market. Who has time to copy a competitor's product and
get it ready in time? Besides, the most you would gain is
one market showing. The guys that specialize at knocking
off successful product don't show at the markets anyway."

Joe looked at Alan. "Are you saying that you're in favor
of this notion?"

Alan shook his head. "No, I'm just saying that I'm not
against it right out of the blocks. I assume the team has
some reason for recommending such a change, and I'm
interested in hearing the advantages. I can think of one
off-hand: we could avoid the mad, last-minute dash to
complete samples for market."

"That's one advantage," Carl said. "Another advantage I
can think of is the ability to control the flow of samples
through the plant, instead of having them disrupt produc-

tion as much as they do at market time."

"Yeah, but what about getting inventory in the warehouse prior to market?" asked Joe. "If we don't plan product introductions around the market, how can we plan to have product in the warehouse when the orders start rolling in?"

Now Frank spoke up. "Joe, you know that we're building samples up until the last minute. Do you really think we have an opportunity to get products in the warehouse? At best, we get a few items in the warehouse in a helter-skelter fashion. But we aren't anywhere close to being able to fill orders. In fact, we're probably in the worst position we could be in. If a suite doesn't sell, we end up eating inventory. If it does sell — we can't fill orders."

"So what great solution do you guys have?" Joe was getting frustrated now. "How would you solve it all?"

Carl placed a hand on Joe's shoulder. "Look Joe, we don't claim to have some kind of magic solution. We're coming at this from a different angle; we're trying to bring up some new ideas. But you're the expert — we want you to consider them and see if they make sense."

Joe took a deep breath. "Okay, so what were you thinking about?"

Carl looked at Lottie. Frank had to admit that he had been surprised at Lottie. She had always been good at what she did, extremely competent. But she had gotten a fire in her eyes since joining the team. He had never seen her so animated. Lottie took the reins from Carl. "Joe, we have been working hard lately to tier our customers, isn't that correct?"

Joe looked confused by the terminology, so Lottie explained. "We've been trying to encourage stronger ties with our good customers by creating 'levels' of dealerships, right? The more items you buy from us, the more discounts, perks, and other benefits you enjoy."

Joe's face cleared up. "Oh yeah, but everybody is doing that. The more you buy from our company, the more we reward the dealer. Call it 'partnering' or whatever you want, but like I said, everyone does it. It's a way of cementing relationships with good customers and trying to encourage lukewarm customers to come on board."

Lottie nodded. "And am I correct in saying that a few dealers do a relatively large percentage of our business?"

Joe nodded. "Right. Like I said, the system is designed to reward dealers who do a lot of business with us."

Lottie smiled. "Okay, then what we're suggesting is this. We introduce new designs in stages. There's a basic 'package' that most dealers buy, as you know, for showroom samples. It's a buffet, a china, a dining table and chairs, maybe a server. The same thing in the bedroom area. This basic package would be introduced to our 'Grade A' dealers. We offer them the basic set at a very attractive price. They get a good deal on a showroom sample, we get a feel for how the units will sell.

"Then we introduce the remainder of the line in two or three stages. Our best dealers get a head start with new lines, and we get a good feeling for how new lines are going to go with our best dealers before we commit a lot of time and materials to the new products."

Joe looked thoughtful for a minute. "Maybe, but I would have some concerns; the fact that the line is limited might hurt sales. We would have to make it clear that the additional units would be coming quickly."

Lottie nodded her head in agreement. "Right. But at the same time being able to work out the details of the line on the base units would allow us to bring the other phases along quicker — we would have shortened the learning curve the second time around."

Frank couldn't resist the chance to add, "Besides, Joe. How many times have I tried to get you guys to drop a piece from a suite because it didn't sell, only to be told we needed to keep it in to round out the line? If it was really that important to have a complete line, don't you think we would sell more of that item? It must not be that important to our customers, if it were many more would buy the complete line."

Joe gave Frank a dirty look but said, "Okay, I'm open to trying this on an upcoming introduction. We'll need to think this thing through, though. I'll want to meet with our sales people and probably some key dealers. We'll have to figure out how to work it in to the partnering agreement. Let me do some work and I'll get back to the team next week."

"Great," said Carl. "You can contact Lottie, she's going to oversee this part of the project."

"Anything else?" asked Joe.

"Well, yes," said Carl. "This item is a little longer term, and frankly we didn't want to open this can of worms until we saw how receptive you were to our first idea. Lottie, you want to cover our other major suggestion?"

Lottie, who had retaken her seat, leaned toward Alan. "Alan, I understand you and Joe have been talking for a long time about using focus groups to test products?"

"Sure," said Alan. "We've wanted to test our new designs and styles before going to market for a long time. We've just never been able to get the time to put together a program for it."

"Well," continued Lottie, "we want to encourage the introduction of a quick and aggressive focus group program."

"Why?" asked Joe. "I mean, I'm not against it, but it

seems to me this is kind of out of your area."

"In a way, it is," agreed Lottie. "It doesn't directly affect what we're trying to accomplish today. However, you know how rapidly we're turning over our product line. In five years less than 10 percent of the products we make today will still be in our offerings.

"In our discussions, we decided it didn't make sense to go to all this trouble and get things in the system now without trying to ensure new products will work well in the system. We want to recommend a series of focus group studies that will enable us to develop a standard set of specifications for our products. For example, we want to standardize our drawer runners, drawer box sizes, door sizes, unit heights, and so on."

"I don't think the designers will go for it," said Joe. "They're probably going to say it would be too restrictive."

"Why?" asked Lottie. "We're not suggesting that they can't be creative, and if a unit absolutely must deviate from the standards, we won't try to kill it. But why not start the process by saying, 'What features are important to our customers, which ones don't they care about, and which ones can we standardize?' Maybe we're being naive, but that's what we are suggesting."

"And how would you go about this?" asked Alan.

"Wes suggested we might build several mock-ups," explained Lottie. "For example, if we wanted to ask customers about drawer box sizes, we might make several drawers and mount them in boxes. Then we load all this stuff up in a rented truck and send somebody to each of our big dealers and some select large customers and get their input.

"No, it wouldn't be perfect, but it would be a damn sight better than just taking a guess, which is what we are doing now," Lottie added. "Wes has agreed to work with product design and sales if you folks are willing to undertake it. We'd like you to think about it; we realize this won't happen tomorrow."

Joe looked thoughtful. "Well, it's an interesting idea. Tell you what, let Alan and I get together and we'll kick around the concept. Ya know, done right, it could even be added as a perk in our partnering program; it would give dealers a chance to influence the design directions of the company.

"In the mean time," Joe asked, "could you put together a list of the types of features you'd want to standardize on? I'll run the list by our two lead designers and see what they think."

Lottie smiled at Joe and Alan. "Thanks, guys. We know we threw a lot at you, and we appreciate your willingness to at least consider our ideas."

Alan smiled. "The entire company looks like it's going to have to change to accommodate this new way of manufacturing. I guess we shouldn't think that we're immune to it. We'll get back to you."

Chapter 18

Employee Involvement in the New Game

So far in our discussion of this new approach to manufacturing we've dealt with specific ideas and concepts on how to improve your business. We've talked about specific ideas, mindsets, and methods to apply to improve your company.

At this point we're going to discuss some decidedly softer issues. The topic is employee involvement — an area that doesn't have a hard and fast set of rules. The ideas laid out here are, in short, my opinions.

Employee involvement is the means by which you involve and motivate the people who work in your factory. This is an extremely nebulous area — techniques and concepts that work wonders in one plant can completely backfire in another.

But there are are a few basic premises that I feel should be a part of every company's effort to give its people ownership of the plant.

1. Stop calling them employees

It may seem gimmicky or only symbolic, but to improve relationships with your workers, stop calling them employees. Call them associates, members, coworkers, or whatever. If you wish, let your people decide what they want to be called. However you do it, try to do away with the negative connotations associated with the "employee" label.

2. Share information

People can't care about what they don't know about. If you've been following the concepts in this book you know that we have been stressing that we need to concentrate on three measurements — Throughput, Inventory, and Operating Expense. At the very least, you should share the information on these measures. However, I favor an open-book form of management — share as much information as possible with your people.

Right about now you're probably thinking what if a disgruntled employee shares that information with my competitors. Unlikely, but say it happens. So what? How many of your competitors are publicly traded companies? You have a wealth of information at your fingertips concerning their sales, costs, even their chief executives' salaries. Does it help you beat them in the marketplace, even if you bother to go through it all?

I remember reading about an incident when a college coach was sent an opponent's play book anonymously. He returned it unopened to the competing school. When asked why, the coach replied, "I have a hard enough time getting my coaches and players to learn the plays in our play book, let alone someone else's."

In short, your time — and your competitors' time — are better spent learning about and improving your own operation, instead of trying to glean some advantage out of information about your competitors. Certainly you should study what your competitors are doing in the marketplace, just as coaches study an opponent's films. It's just not likely that running an open-book style of management will come back to haunt you, and the advantages gained from treating your people like adults far outweigh the risks.

3. Use teams

The use of manufacturing teams in your company will open a host of opportunities, but don't go overboard. I have been fortunate to work in environments where teams played a big part in the turnaround of a company. The changes in the people were nothing short of miraculous. It wouldn't be an exaggeration to say that, in at least one case, team spirit saved the plant.

However, I've also seen teams become overused. Anytime we had any problems in the plant, we threw a team at it. I also saw some people forced into the team leader's role who, to put it bluntly, weren't able to function in that capacity. It was done for a good reason: to ensure that everyone on the team went through being a leader. But in the end, I wonder if the team didn't suffer more than it was helped.

Even where you don't use any formal teams, you still want to foster a team spirit. It should become part of the whole mindset we've been considering including universal measures, common goals, and everyone working as a team.

4. Incentive systems can help

Few subjects in the area of motivation spark as much discussion and controversy as the topic of incentive systems. When done correctly, incentive systems can be an important contributor to your company's success. Here are a few observations on the subject that should prove helpful.

• Piecework or individual incentive systems don't work in flow manufacturing environments. These type of systems have their defenders, but incentives tend to reward people who figure out how to beat the system, rather than those who contribute the most to the company. They encourage workers to pass on substandard parts and accept inferior quality for the sake of "making the rate."

For the same reason, I don't like departmental incentives. They encourage departments to work against each other instead of with each other to achieve a common goal.

• Feedback has to be consistent, continual, and delivered at the individual, team, and plant levels. While individual incentives can be counterproductive, feedback provides an excellent way to review and improve performance. Showing you care enough to talk about a person's performance with them can be a powerful incentive in itself.

• If an incentive system is to be used, it should be on a plant-wide basis and be based on Throughput, Operating Expense, and Inventory (T, I, and OE). I'm not a fan of incentive systems or gainsharing that try to incorporate too many criteria.

I have no problem with gainsharing, as long as it's based on T, I, and OE. Problems always seem to crop up when you start trying to throw in additional items such as quality, safety (why in the world we need to offer incentives not to be injured is a mystery to me), attendance, and other human resource issues.

A Better Approach

If you want to reinforce the idea that quality and safety

Employee's Profit-Sharing Report

Benchmark item	Total amount	Amount per employee
Gross profit sharing amount	$300,000	$600
Amount you lost due to bad quality (returns, credits, repairs)	(75,000)	(150)
Amount you lost due to poor safety (claims, disability payments)	(62,500)	(125)
Total profit sharing this month	$162,600	$325

have an impact on the bottom line, you could use a monthly report form such as the one shown here for a company with 500 employees.

Of course, you also could detail any aspects of Throughput, Inventory, and Operating Expense (such as overtime) you wanted to highlight in this breakdown. This provides a way to encourage the behavior you want without building a lot of factors into the profit sharing system. If the behavior is really that important, it's going to affect Throughput, Inventory, and Operating Expense.

In today's competitive environment, we want more of our people than ever before, but we must give something in return. We have to treat those who work with us as adults, giving them the same respect and consideration we ask for in return. It must truly be a team effort if we hope to win the game.

Chapter 19

A New Beginning

Frank and Wes were huddled together in Frank's office. They were working on a proposal for the company's incentive system. When Wes had come on board, they had been right in the middle of implementing a department incentive system. Wes had worked quickly to stop that process, but had promised to develop a new system with Frank A.S.A.P. They were putting the finishing touches on their proposal.

"So we'll use the rates only on a job-by-job basis, and use them only for team feedback, not as a measurement for the incentive system?" asked Frank.

"Right," said Wes. "If an operation is supposed to take one hour to run, we want the operators to try and get that operation done in one hour. What we don't want to do is encourage them to stay busy for appearance's sake. So, if they only get that one load today, that's fine.

"But when they have work in front of them we want them to try and finish it within the standard time. People still need benchmarks to judge their performance, and to let their leaders know how they're performing. We simply aren't going to let the system dictate the production of parts that aren't needed just to get the measurements up."

"But what about concepts like seniority, pay for knowledge, and other incentives?" asked Frank.

Wes shook his head. "Look, if your company wants to build in something like pay-for-knowledge, it will have to be done at the base wage level. This incentive system is intended to do one thing and one thing only — help the company make more money.

"Now, it just so happens that if you do a better job on quality and rejects go down, the company makes more money. It just so happens that if you work safely, and compensation costs go down, the company makes more money.

"Profitability should encourage a lot of positive things we want to happen, and will if we present the data in an effective manner. But in the end we have to remember why the incentive system is here, and it's the same reason we are here — to make more money."

Wes finished writing a few lines, then stretched. He handed the finished sheets to Frank. "Read over those last few paragraphs, Frank, and if you agree, we'll get it typed up, and you can send me a copy."

As Frank read the plan, Wes got up and started gathering up his things. He put his coat on and came over and stood by the table, watching Frank read.

Frank shook his head. "I'm sorry Wes, I think it still needs work."

Wes sighed and started to slip off his coat. Frank laughed and said, "I'm kidding, I'm kidding! I think it's fine, and I know you have a plane to catch."

Frank walked Wes to the parking lot and helped him load the rental car with his luggage. "I'll be calling either you or Carl every week, and I'll be back in two months to check on things."

Frank shook Wes' hand. "I want to thank you, Wes. I'll

admit I had my doubts. We've still got a long way to go, but I feel we at least can see the path we're headed down now. The increase in throughput and reduction in inventory have been nothing short of miraculous to me, even though I have been here for every minute of it."

"You and your people deserve the credit, Frank," Wes said. "Most of the answers to problems along the way came from you and your folks. All I did was get you thinking in a different way about why you were here. All of you had the tools for success all along. You just needed a better way to use them."

Frank watched Wes drive away, then started walking back to the office. As he walked by his father's portrait in the lobby he made a mental note to call and invite him down to the plant. Frank smiled. He was pretty sure his father would be happy with what he saw.

Chapter 20

Not the Last Word

In this book, and the series of articles that preceded it, we've attempted to introduce you to a different way of thinking about your job and your company. That's what synchronous manufacturing, flow manufacturing, and similar concepts have to offer at their most basic level — a different way to look at what you face every day.

Lots of books have claimed to have THE ANSWER — the solution to all your manufacturing problems. Why should you really believe that the solutions suggested here are any different? In the end, of course, you must make that determination for yourself.

Like you, I had to evaluate and choose between traditional manufacturing on the one hand and MRP, J-I-T, Kaizen, TQM, and all the other manufacturing strategies that have come out over the years on the other. I can tell you why I believe that flow manufacturing is the best solution to today's manufacturing problems.

To start, flow manufacturing doesn't try to invalidate the other concepts. Indeed, we make extensive use of these ideas during a project. This book has offered numerous examples of where we used ideas such as setup reduction and Poka-Yoke. All of these concepts are acceptable with and readily adaptable to flow manufacturing, which views them all as potential tools to achieve the basic goal — making more money.

For another, these concepts have been constantly evolving. At various times, they've been known as the Theory of Constraints, Synchronous Manufacturing, Flow Manufacturing, and other names. I feel sure the methodologies we use five years from now will be different from today's as we constantly refine the concepts we talk about. In other words, this is a system that recognizes the need to adapt and grow, unlike other concepts that were the "last and final word" on how to run a plant.

Finally, we don't present this as a cookie-cutter solution. The application and implementation changes with each company, depending on its needs and the specific conditions of the project.

In the end you must judge for yourself. I do hope that this book has caused you to reevaluate your company and shown you how to go about achieving your revised goals. If so, then my personal goal in writing this book has been achieved. I wish you and your company great success!

Afterward

THE W.E.R.C. SYSTEM

The W.E.R.C. (or Woodworking Equipment Reduction of Changeovers) system is one I developed while working for HON Furniture. It's based on the original work done by Shigeo Shingo, who developed many of the J-I-T concepts used at Toyota. However, the original setup reduction programs were aimed at large presses and similar machines that had setup times between two and eight hours. My seminars are directed more toward the equipment used in woodworking.

Although the concepts necessary for a complete understanding of W.E.R.C. are usually taught in an eight-hour or three-day seminar format, the basics of the approach are as follows.

STEP 1: VIDEOTAPE THE SETUP AND BREAK IT DOWN INTO ELEMENTS

The basic setup reduction process selects a typical

setup operation on the machine you wish to improve and videotapes a complete changeover. A changeover is defined as the total time between the last good part of one run and the first good piece of the next run — so you need to videotape everything involved.

Next, the setup reduction team goes through the setup operation and breaks it down into steps, noting the time each step takes. It's useful to summarize this information on a bar graph, as it visually shows which steps take up the most time.

STEP 2: IDENTIFY ELEMENTS
AS INTERNAL OR EXTERNAL

Once the setup is broken down and timed, each step should be classified as internal or external. This is an important step in reducing machine downtime between runs. Internal operations are items that have to be done while the machine is off. For example, if you were working on a tenoner, changing the saw blade is something that has to be done when the machine isn't operating.

External operations, on the other hand, are items that *could* be done when the machine is not running. On the videotape they might be done when the machine is off, but the team should ask itself whether this step could be done while the machine is producing. Some examples of these types of items are paperwork or parts inspections.

STEP 3: CONVERT INTERNAL ELEMENTS
INTO EXTERNAL ELEMENTS

As you do this analysis, you'll be surprised at how few of the steps really have to be done while the machine is

off. With this knowledge the team should work at con-
verting internal operations to external operations. If parts
have to be inspected, have someone else inspect them
while the machine is running. If tooling needs to be trans-
ported to and from the machine, have it done by someone
else while the machine is running. Many steps can be
converted from internal elements into external elements
when you use this approach.

STEP 4: STREAMLINE INTERNAL ELEMENTS

Here you reduce the time of the remaining internal elements. Some ways to do this include:

1. Have all tools on hand and identified. Don't share
tools between different machines.
2. Use locating pins to put tools and fixtures in their
proper place quickly. Don't rely on measuring to locate
items.
3. Don't use bolts that require more than one turn to
tighten.
4. Don't use wood blocks as guides/fences.
5. Make all jigs and fixtures the same height, same
locating point, etc.
6. Don't have different size bolts, nuts, and other
connectors.
7. Use multiple people on the setup in the same
manner that a racing car pit crew operates.

STEP 5: STREAMLINE EXTERNAL ELEMENTS

Use the same ideas as presented above to reduce the
external time of the setup.

STEP 6: PRACTICE THE REVISED SETUP

These, in the most basic terms, are the steps you can take to reduce setup times.

About the MPI Group

The MPI Group is a resultant firm (we prefer not to use the "C" word) that specializes in improving a company's profitability through the application of Synchronous Manufacturing or Flow Manufacturing Techniques.

All of MPI's Associates are veterans of the shop floor with backgrounds in furniture, metalworking, extrusion, electronics, light and heavy assembly, service — virtually any sector of business and industry. The MPI Associates have more than 100 years of collective experience in implementing these concepts.

MPI puts on free seminars across the country to introduce its concepts and Associates to prospective clients. In addition, it offers plant walk-throughs and in-house seminars upon request. The requesting company is required only to pay for travel expenses.

What type of results can a typical company expect from a project?:

Inventory reductions of 32 percent.

Lead time reductions of 50 percent.

On time deliveries increase to 95+ percent.

All adding up to an increase in bottom line profits.

Terry Acord heads up MPI's Carolina Office, serving the

Virginias and the Carolinas. You can contact him by phone at 919-742-2111, by fax at 919-742-3959, or by E-mail at tdacord@mail.emji.net

To request more information on the range of services offered by the MPI Group or to schedule a plant walk-through contact Mitch Slater at MPI's Connecticut office; phone 203-294-1940 ext. 704 or fax 203-294-1949.

Printed in the United States
3620

9 781574 500509